Record-Breaking Bugs

Amazing Insects

Written by Matt Turner
Illustrated by Santiago Calle

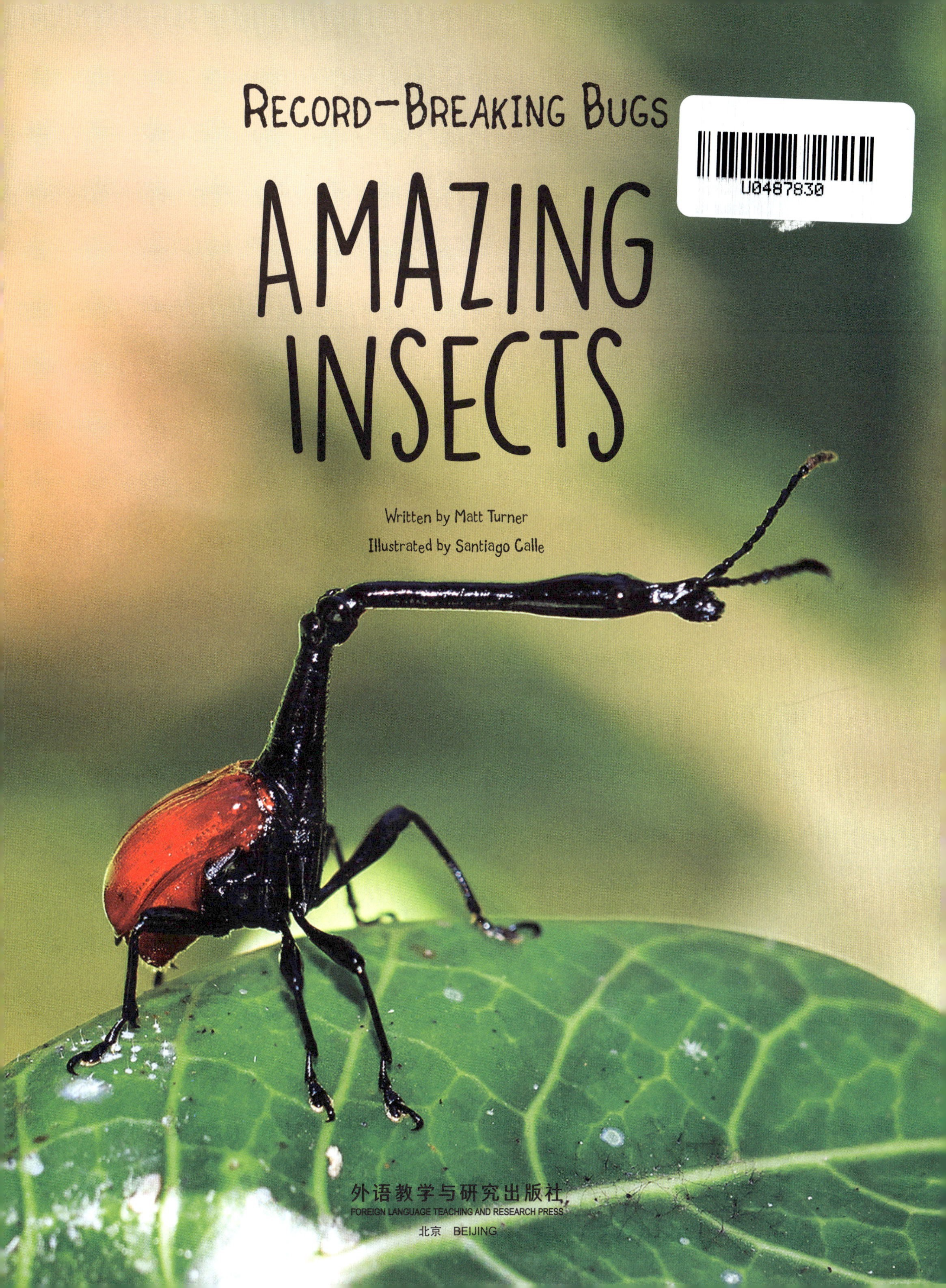

外语教学与研究出版社
FOREIGN LANGUAGE TEACHING AND RESEARCH PRESS
北京 BEIJING

There are about ten quintillion (10,000,000,000,000,000,000) insects alive at any one time – they have been around for nearly 500 million years! So step inside and get close to Earth's great survivors, many of whom live in armies and colonies. You'll find that

- all the world's ants weigh more than the world's people
- the mantis looks exactly like a flower or a rotten leaf
- some termite queens lay up to 40,000 eggs a day

  ...

and much more. You will also read stories of ant armies, leaf insects (first thought to be actual walking leaves), tower-building termites and flashing fireflies. Insects rule, OK?

Contents

EXTRAORDINARY INSECTS

We humans may think we rule the planet, but actually the insects are in charge!

For a start, they far outnumber us. At a rough guess there are about ten quintillion (10,000,000,000,000,000,000) insects alive at any one time, made up of more than 900,000 species. And they're far older than us, going back nearly 500 million years, long before the dinosaur age. They're about 120 million years older than the flowering plants, too. In other words, flowering plants evolved to survive alongside existing insects, rather than the other way around.

Some insects live alone, but the more advanced – ants, bees, termites – form complex societies, working together to build megastructures, gather food, nurse their young, and defeat enemies.

This book looks at some of the most amazing adaptations of modern insects: their beauty, camouflage, engineering skills, senses, weapons, defences and sneaky survival tricks, plus some of the ways – both welcome and otherwise – they affect humanity. And the great thing about them is you have an insect zoo in your house and garden. Go take a look…

GREAT SURVIVORS

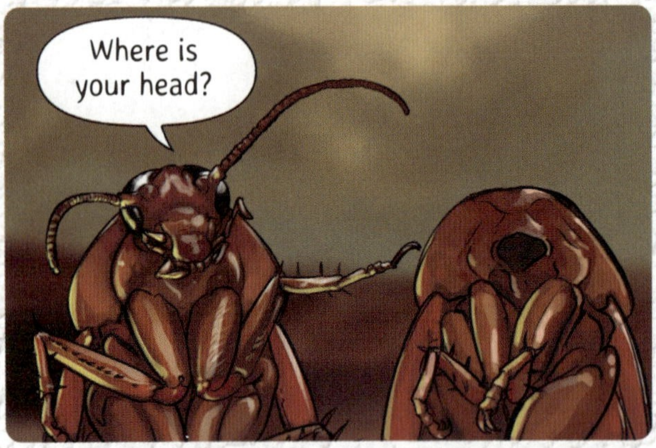

A cockroach can live for weeks without its head. And the severed head can keep on waving its antennae for hours!

Cockroaches first appeared about 320 million years ago. Modern-day cockroaches developed about 200 million years ago and walked with the dinosaurs.

Cockroaches are scavengers. In a pinch they will eat just about anything: glue, grease, soap, wallpaper paste, leather or even hair.

Some cockroaches are big. *Megaloblatta longipennis* boasts a 17.5-cm wingspan; the Australian rhino cockroach can weigh 33.5 g.

Cockroaches are fast! Once all six legs are in motion, a cockroach can sprint at speeds of up to 1.5 m per second. And they're elusive, too, with the ability to 'turn on a dime' while in full stride.

COCKROACHES

Flick on the kitchen light at night, and you may see a flat brown insect racing over the floor to a dark hiding-place. Few insects are so unloved as the cockroach – but, of about 4,600 known species, only 30 or so are pests. Some are specially adapted for cold or dry habitats, but most are generalised, eat-anything, go-anywhere critters… which helps explain how cockroaches are found in almost every corner of the planet.

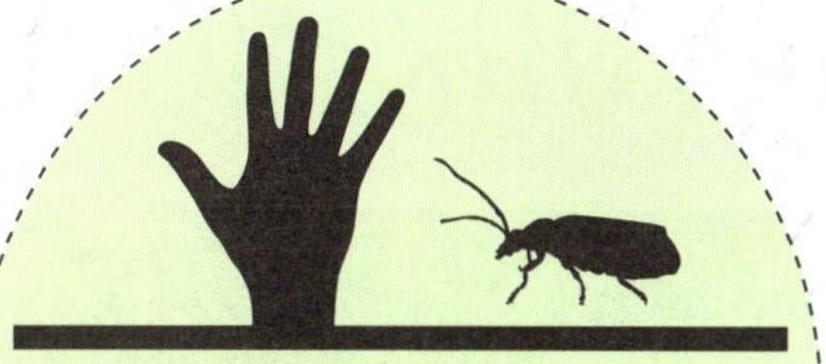

MADAGASCAR HISSING COCKROACH

Body length: up to 76 mm

Lifespan: 2–5 years

NUCLEAR PROOF

In a nuclear attack, among the survivors would be fruit flies and cockroaches. They can handle mega-doses of radiation that would kill humans.

SWEET TOOTH

If a cockroach is introduced to a sweet taste, such as vanilla or peppermint, it will drool in anticipation when it later detects the scent in the air.

Working Together

Ants of all sizes cooperate. Left: An *Atta* major carries a leaf fragment, giving minors a ride; in return, they guard her from parasitic flies. Centre: A *Pheidologeton* supermajor acts as a 'troop transport' for minors. Right: An army worker grooms a soldier's jaws.

A weaver ant gently holds a larva and taps it with its antennae to make it release silk. The ant uses the silk to stick leaves together and build a colony home.

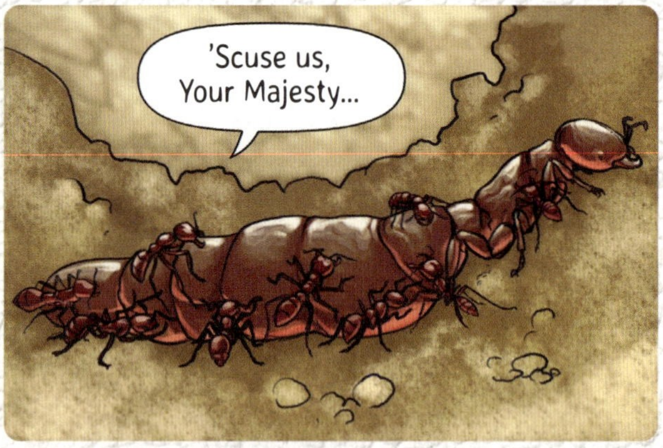

A queen driver ant is so big – up to 5 cm long – that her tiny workers, just one-tenth her size, have to push her around.

Some workers in carpenter ant colonies take defence to the limit. When attacked, they explode and die, blasting toxic gunk all over their enemies.

Honey ant workers, or repletes, hang from the ceiling of the nest and regurgitate liquid sugar and protein from their abdomen to feed the other ants.

ANTS

Ants are found worldwide, having evolved only about 130 million years ago from ancient wasps. They form colonies, sometimes millions strong, made up of several castes – queen, worker, soldier, and so on – all performing different roles. They work together to build the nest, raise the young and fend off enemies. Ants will fight to the death for the colony's survival.

BULLDOG ANT

Body length: 8–40 mm

Lifespan: up to 2 years approx.

FIERCE

A bulldog ant is so fierce that, if it is cut in two, the head will still try to bite the abdomen, and the abdomen will sting the head.

HEAVY

If you took a giant set of scales to the Amazon rainforest and put all the ants on one side, and all the land vertebrates (mammals, birds, reptiles and amphibians) on the other, the ants would be around four times heavier.

Fearsome Larvae

Antlions undergo complete metamorphosis: they go through a complete egg-larva-pupa-adult life cycle. First, a female adult antlion deposits eggs in dirt or sand.

The hatched larva digs a cone-shaped pit, burying itself at the bottom, but leaving its long, 'toothed' jaws free. Then it lies in wait.

If prey tumbles in, the antlion may flick sand up to cause a 'sandslide', knocking the victim off its feet. Bursting out from its hiding-place, the antlion uses the hollow 'teeth' of its jaws first to inject its victim with digestive venom, then to suck out the resulting 'soup', leaving nothing but an empty husk.

The larva pupates inside a round cocoon about the size of a chickpea. It may remain here, buried in sand, for several years.

Finally, the long, slender, winged adult emerges from the pupa. It spends the next few weeks searching for a mate, and not feeding at all.

ANTLIONS

Antlions are not ants, with about 2,000 species found worldwide. While the adult is a winged insect that looks like a rather scrawny dragonfly, the larva is a more fearsome beast. And while sometimes it simply hides among leaves or in rock cracks, it is best known for its habit of digging a sandpit, where it lurks with jaws gaping wide, waiting to capture prey and suck it dry.

ANTLION

Body length: larva up to 15 mm, adult up to 80 mm

Lifespan: larva 3 years or more, adult 25–45 days

SNEAKY

The name 'antlion' is centuries old. Maurus, a medieval scholar, said of the larva that 'it conceals itself in dust and kills ants that carry provisions'.

NICKNAME

Some Americans call antlions 'doodlebugs'. In *The Adventures of Tom Sawyer*, the classic tale by Mark Twain, Tom talks to a doodlebug to coax it out of its hole.

Borers and Battlers

These are male giraffe weevils from New Zealand. They look rather different from the Madagascar giraffe weevil opposite, but they probably fight the same 'snout battles' to decide who mates with females. Their huge snouts measure up to half the male's total length of 9 cm.

In the acorn weevil it's the female who has the longer snout. She uses it to bore a hole in an acorn, then turns around and lays an egg in it.

The larva develops inside the acorn, then drills its way out and drops into the soil, where it pupates for a year or two before becoming an adult.

The palm weevil lays up to 500 eggs at a time in coconut, date and oil palms. Its big larvae chew through the timber, ruining the crops.

Insects are rich in protein. In Malaysia, people cook the weevil larvae in a dish called 'Sago Delight'. In Vietnam, weevils dipped in fish sauce are eaten alive. Mmm!

WEEVILS

Weevils are smallish beetles, usually with a long snout, and their antennae have an ‘elbow’ joint. There are over 60,000 known species, and although farmers hate them because they burrow into plants and damage crops, weevils have some fascinating adaptations. In some species the snout – or rostrum – reaches extraordinary lengths, and males use theirs to battle each other over females.

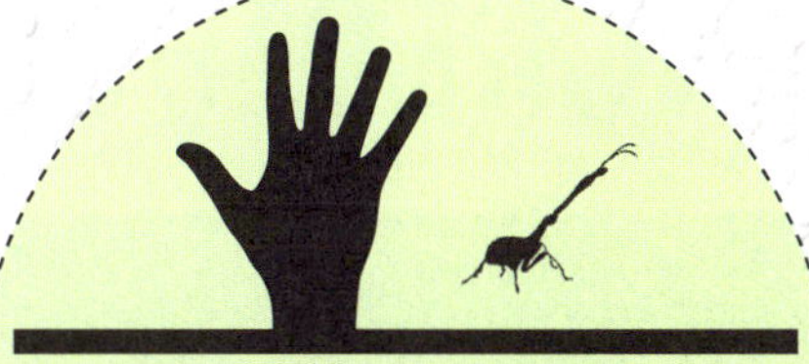

MADAGASCAR GIRAFFE WEEVIL

Body length: up to 25 mm
Lifespan: 1 year approx.

QUANTITY

Nearly one in four of all life forms are beetles – and of these, around one in five are weevils. Weevils are almost everywhere!

WELL-NAMED

The male Madagascar giraffe weevil is nearly all neck! After mating, the female (which has a much shorter neck) carefully lays her eggs in rolled-up leaves, one egg per leaf.

INCREDIBLE HULKS

Giant weta regularly weigh 35 g, more than a mouse. The heaviest on record – a pregnant female – tipped the scales at 70 g.

Like her cousins the crickets, the female weta has a spiked ovipositor, which she uses for laying eggs. She pushes the eggs into damp soil.

The mountain weta lives high above the snowline. It can survive being frozen stiff! It just thaws out in warmer weather... and carries on.

Tusked weta were discovered in 1970. Males use their tusks in jousting battles over females, and rub them together to make a rasping noise.

Cave weta aren't giants, but they do have long legs, which may span up to 21 cm on a body just about 3.5 cm long. You can find lots of them clinging to the roofs of cool, damp caves in New Zealand, so if you go looking for them, remember to check your coat collar afterwards! They're good at jumping, too.

GIANT WETA

A flightless vegetarian with a face like a samurai warrior, armed with powerful jaws and spiky legs… and heavier than a sparrow? Meet the giant weta! All 11 species live in New Zealand, where rats, stoats and other introduced predators have wiped out their mainland populations, leaving the only survivors on islands or reserves. There are smaller weta species, too, living in caves, forests and even high mountains.

GIANT WETA

Body length: up to 10 cm
Weight: up to 70 g
Lifespan: about 2 years

ISLAND GIANTS

Giant weta are examples of island gigantism: animals or plants that evolve in isolation on islands can, over time, become enormous.

UGLY

The Maori people of New Zealand named this insect *wetapunga*, after the god of ugly things, meaning 'ghost' or 'spook'.

Leaf Lookalikes

Antonio Pigafetta found leaf insects when he explored the Philippines in 1519–1522. He thought they were 'walking leaves' that fed on air.

Nymphs shed their skin several times as they grow. If they lose a leg or antenna, they can regrow it – the limb gets bigger with each moult.

Phasmids have amazing night vision. With each moult, the eyes grow bigger, but more sensitive. So while nymphs come out by day, adults are mostly nocturnal.

Some leaf insects spray a defensive fluid from glands in their neck, to stop predators bothering them. If it gets in your eyes, it's very painful.

Relatives of the leaf insects include the two-striped walkingstick or 'devil rider', an American species. The male and female are usually found attached together in summer and autumn when mating. They go everywhere together and they don't separate until one of the pair dies and drops off!

LEAF INSECTS

The leaf insects or 'walking leaves' have taken camouflage to the absolute limit, with bodies that have evolved to look exactly like real leaves. They have veins and ribs, and even crinkly brown edges. Related to stick insects, they are known as phasmids. Like some wasp species, leaf insect females can produce eggs without the help of a male, and the eggs hatch… into more females!

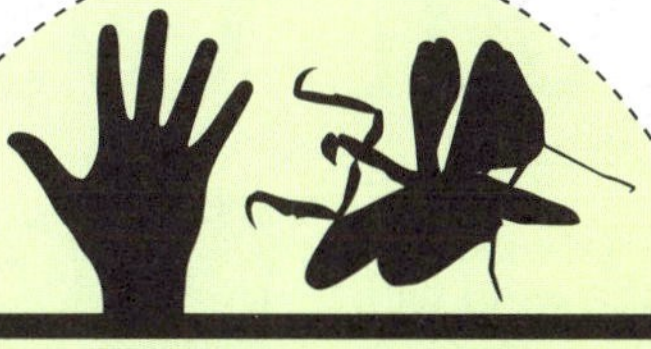

GIANT LEAF INSECT
Body length: up to 11 cm
Lifespan: 5–7 months

RARE MALE
There are male leaf insects, but they're incredibly rare. The first male giant leaf insect was only discovered as late as 1994.

CANNIBAL
Stick insects make great pets. But if they run out of leaves to eat, they'll eat leaf insects instead… so you can't keep two together in one cage!

Cool Camouflage

The mantis is the only insect able to move its head through 180° each way. So it can perch stock-still and keep watching prey carefully.

The two dots that 'float' around the eyes are false pupils: the part of each eye that absorbs light while the rest of the eye is reflecting it.

Like the leaf insects, mantises walk in a wobbly way. They do this for an extra camouflage effect, swaying like a twig in a breeze. The flower mantis on the left, however, is putting on a deimatic display – a startling move that shows off its bright colours. It sometimes does this to scare off predators.

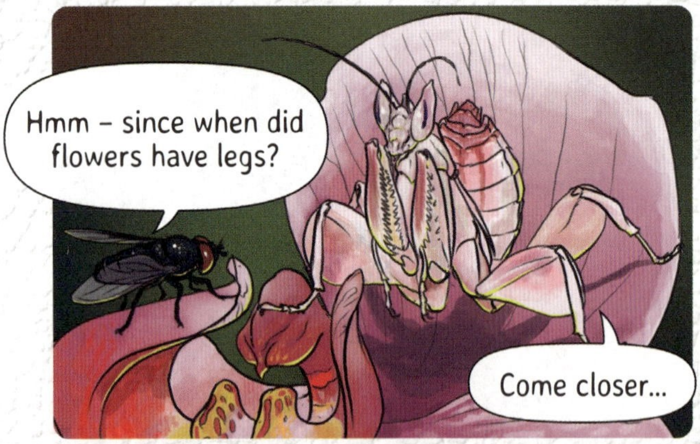

Insect camouflage reaches beautiful extremes in species like the orchid mantises. They're hard to pick out from the flowers on which they lurk!

The mantis egg sac looks like a squeeze of toothpaste, often fastened to a garden fence. Newly hatched nymphs look just like mini mantises.

MANTISES

The mantis is an ambush expert, a predator so well-camouflaged it melts into the scenery when perched on a plant. The big compound eyes watch for movement, and if another insect settles near, the mantis flicks out its long, spiked forelegs to snatch it. The many different species show wonderful variations of camouflage – some look like colourful flowers, others like rotten leaves or dry twigs.

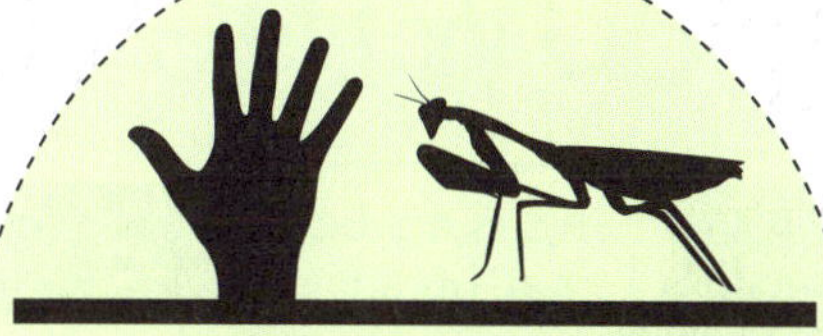

VIOLIN MANTIS

Body length: female 10 cm, male 8–9 cm
Lifespan: 1 year, sometimes 2

THORNY

The violin mantis has an extra-long thorax, which resembles the slender neck of a fiddle. It also looks just like the dry twigs of rose bushes, where it likes to hide.

VANISHING ACT

A young mantis may change colour – from brown to green, say – when moulting, in order to blend more closely with its environment.

Glowing in the Dark

A male *Photinus* firefly flashes a special pattern to attract a female of the same species, who then flashes a 'Come here!' answer that leads him to her.

Sometimes, a sneaky female from the genus *Photuris* tricks a *Photinus* male by flashing the correct answer... and then eats him when he finds her!

The New Zealand glow-worm is the cave-dwelling larva of a fly. It lives in a 'sleeping bag' of mucus, slung from the roof on lines of silk. The larva spins silken threads up to 50 cm long, covered with beads of sticky mucus. Then it 'switches on' the bioluminescent light at its back end...

Any flying insect that bumps into a dangling thread is instantly trapped by the mucus; now the glow-worm larva reels it up and eats it.

In America, some beetle larvae glow to lure prey, such as millipedes. They are known as 'railroad worms' because they look like a train lit up at night.

GLOW-WORMS and FIREFLIES

Glow-worms and fireflies can give out light, like a living torch. There are over 2,000 species: some shine to warn predators they taste bad, while others do it to attract a mate, or to lure prey. The natural light, or bioluminescence, comes from a chemical reaction in the abdomen. It's similar to what happens when you crack a glow-stick – except that the glow-worm can keep doing it.

COMMON GLOW-WORM

Body length: larva 3–25 mm, adult 15–30 mm
Lifespan: larva 2–2.5 years, adult 10–16 days

EFFICIENT

Glow-worm bioluminescence is a 'cold' light. Almost 100 per cent of the energy used is converted to light, with none wasted as heat.

LIGHT DISPLAY

Sometimes all the fireflies in one area will flash on and off at the same time, in a huge synchronised display. No one (except the fireflies) is quite sure why.

Beetle Defences

The bombardier beetle mixes chemicals in its abdomen to blast out a shot of scalding-hot poison, harming or even killing would-be predators.

Ladybirds ooze haemolymph (blood) from their limb joints, which not only tastes bad but also hardens on air contact, gluing up the jaws of attackers.

The palmetto tortoise beetle larva (left) hides in a 'tent' made from strings of its own poo. The adult beetle (centre) has big feet, each with about 10,000 bristles and glands that can ooze a gluey oil. When threatened, it glues its feet to the floor (such as a leaf) and becomes impossible to dislodge (right).

If it crash-lands on water, the rove beetle *Stenus comma* skates away using 'fart power'. Gases released from special glands jet-propel it to a safe place.

Carrion beetles (and their larvae) feed on decaying animal carcasses – really dirty places! They give off ammonia to keep free of harmful bacteria.

BEETLES

There are more species of beetle than of any other animal group on Earth. Many are eaten as prey, turning up on the menu of birds, lizards, mammals and spiders, so they've evolved an arsenal of spectacular defences, including deadly poisons, explosions, cleaning fluids and jet propulsion. Even the humble garden ladybird can squeeze out something nasty to defend itself!

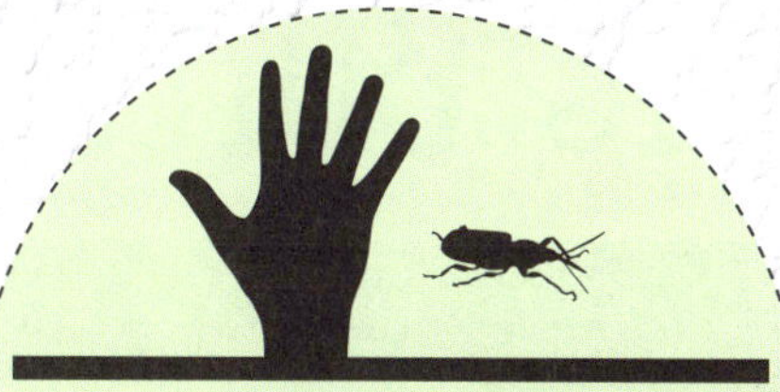

ROVE BEETLE
Body length: up to 40 mm
Lifespan: adult 20–60 days

EVOLUTION
Chemical defences of one kind or another have evolved more than 30 times during the 270-million-year history of beetles.

TOXIC
The haemolymph of some rove beetles contains pederin, a super-deadly toxin that is thought to deter predatory spiders.

Big Builders

Welcome to termite society! The queen, tended by a king, spends her long life laying eggs. Workers and soldiers are blind and cannot reproduce. Every so often, the queen produces some alates: these are winged termites that breed to start new colonies. In the picture above, can you work out who's who?

Termites are architects and farmers! Some termites build huge mud towers, complete with basements, air-conditioning ducts and fungus farms.

Soldiers defend the colony. Some have jaws, but this is a nasute: a soldier with a nozzle-like head for squirting toxic goo at ants and other enemies.

Drywood termites are a pest, living in our walls, chewing the timbers and causing billions of dollars' worth of damage worldwide each year.

TERMITES

Termites are tiny social insects with a huge impact. Some species ruin houses and crops, but many others are the farmer's friend, improving the soil surrounding their massive colonies, which contain thousands, even millions of workers and soldiers, all slaving together in blind obedience to the queen (pictured below). And what a queen! Measuring up to 10 cm long, she may live to be 50 years old – an insect record.

TERMITE

Body length: workers 4–15 mm, queen up to 100 mm or more
Lifespan: workers and soldiers 1–2 years, queen up to 50 years

MINI-COWS

Termites have been referred to as 'mini-cows' because their multi-chambered gut can break down cellulose, the tough stuff in plants.

FOOD CHAIN

Winged termites provide more food for the mammals, birds and amphibians of tropical forests than any other insect group.

Record-breaking Insects

Insects may be small, but they notch up some fascinating records. You won't find the biggest, deadliest, fastest insects on these two pages: instead, here are some feats that relate to the book. Also, you can study some of these insects in the house and garden – although hopefully there are no termites hiding in your walls!

Night light: The brightest bioluminescent insect is the Jamaican click beetle. The light comes from two 'headlights' on the top of its thorax, as well as a glowing patch on the underside of the abdomen. The light is so bright that local people used to collect the beetles in gourds and hang them from the ceiling to illuminate their huts.

Changeling: The insect that undergoes the greatest number of moults is the silverfish. This ancient creature, little changed since the dinosaur era, gets into our homes and nibbles our books and clothes. It may shed its exoskeleton 60 times or more.

Supercolony: The largest ant colony found to date was in Southern Europe, made up of 30 linked colonies across an area 6,000 km wide. It contained untold billions of workers. Incredibly, there is also a 'megacolony' of Argentine ants living in far-flung lands, from California to Europe and Japan. Ants from different groups will usually fight, but if they belong to the same supercolony, they act friendly when brought together, proving their kinship.

A rovin': Talking of big families... the largest family of beetles is the Staphylinidae or rove beetle family. It contains more than 62,000 named species, and there are plenty more to discover. In fact, there are more species of rove beetle than of all the vertebrates (animals with backbones) put together.

High-rise: The tallest insect-built structures, relative to insect body length, are the termite mounds in Africa, some measuring up to 9 m tall. In human terms, that's like a mile-high skyscraper. What's more, termite nests can extend deep below ground... to more than 60 m in some cases!

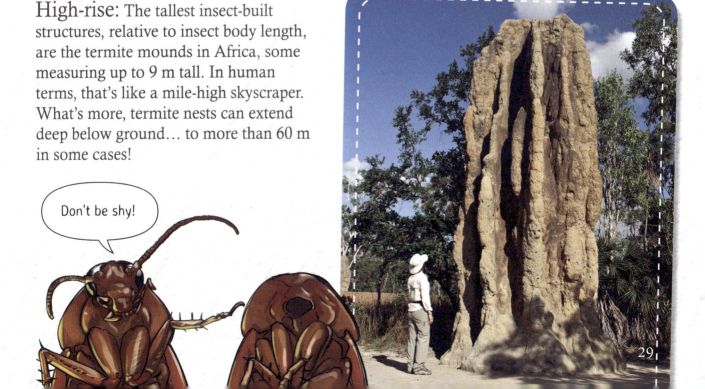

Amazing Insect Facts

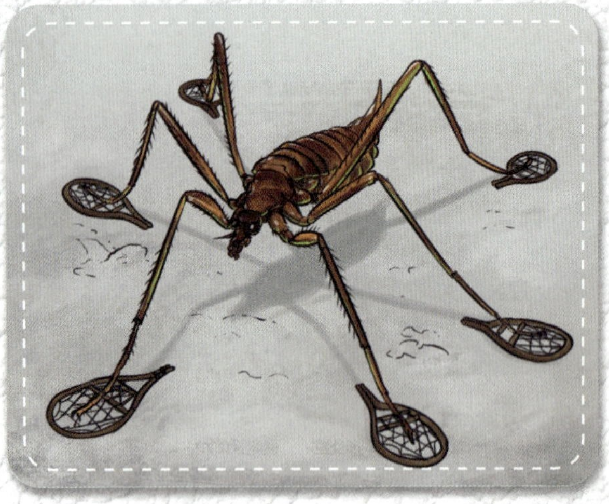

What do you call a fly with no wings?

A walk! Actually, there are lots of wingless flies. They include the snow flies in the genus *Chionea*, which – as their name suggests – can be found in winter, walking on snow. Their haemolymph contains a natural antifreeze. Other wingless examples include bat flies – true flies that live as parasites on the fur of bats, and can only take flight by hitching a ride.

Can insect toxins save human lives?

Scientists are studying several insect toxins for use in anti-cancer drugs. These toxins include pederin (see page 25), as well as DHMA, a rare fatty acid found in soldier beetles. Another is MP1, a toxin from the venom of a Brazilian wasp, which seems to kill cancer cells while leaving normal cells unharmed.

How many ants are there?

At any one time, there are about a quadrillion of them – 1,000,000,000,000,000 – on Earth.

Can goats rescue insects?

Introduced rats, stoats and other predators killed so many giant weta in New Zealand that the insects were believed extinct on the mainland. Then in 1962, giant weta were discovered in a clump of spiky gorse bushes near Mahoenui in the North Island. They had survived because the grazing of goats had kept the gorse growth so dense that rats couldn't penetrate it. More than 50 years on, the weta are still there.

Glossary

abdomen the hind part of an insect's three-part body (along with the head and thorax).

antenna (plural: antennae) one of the two slender, movable sensory organs on an insect's head, providing touch, taste and smell.

caste in insects that form societies, such as ants, bees, wasps and termites, a caste is a social group with a particular function, such as a worker or soldier.

cocoon in insects that undergo complete metamorphosis, the cocoon is the casing around a pupa, inside which an insect larva gradually transforms into an adult.

gland an organ that can produce a chemical and release it, either into the body or outside it.

introduced predator a predatory animal brought (intentionally or otherwise) by humans from its native range into a foreign range. Introduced predators often upset the local ecosystem.

metamorphosis the process in which an animal, such as a butterfly, goes through a number of life stages, including egg, larva (nymph, for the insect undergoing incomplete metamorphosis), pupa, adult. Hemimetabolous insects skip the pupal stage.

moult in insects, to shed the old exoskeleton (outer part) in order to grow. The new exoskeleton is soft but hardens after moulting.

ovipositor a hollow, often spike-like organ on a female insect's abdomen, which she uses to lay eggs.

predator an animal that kills and eats other animals. Roughly one-third of all insect species are predatory.

regurgitate to 'sick up' digested food from the stomach. Some ants often feed one another on regurgitated food.

rostrum the snout or beak of an insect, such as a beetle.

scavenger an animal that feeds on whatever it can find, such as fallen fruit, carrion (dead animal remains) or garbage.

thorax the middle part of an insect's three-part body (between the head and abdomen).

venom poison that is injected (from an ant's stinger, for example) into prey or an enemy.

vertebrate an animal with a backbone, such as a mammal, bird, fish, reptile or amphibian.

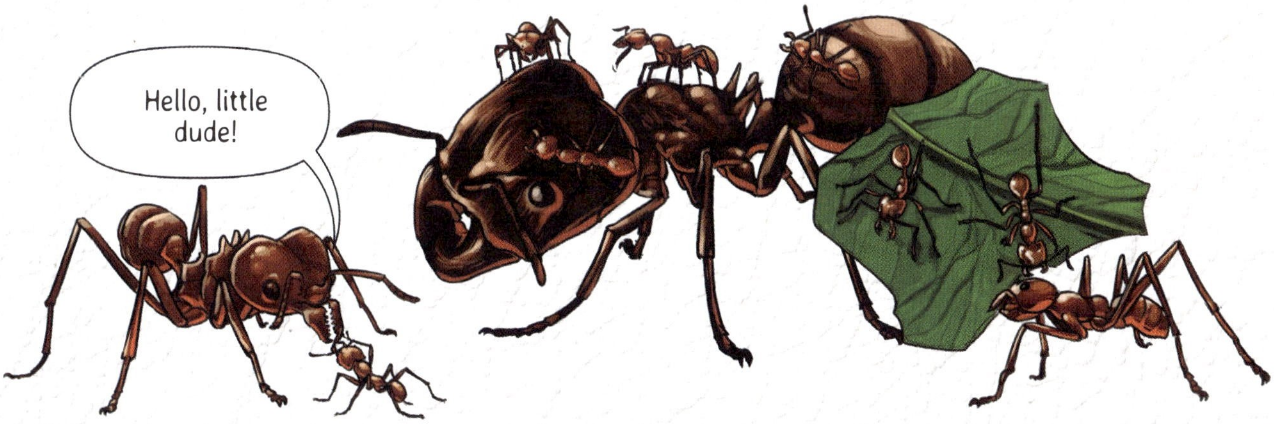

INDEX

The Author

British-born Matt Turner graduated from Loughborough College of Art in the 1980s, since which he has worked as a picture researcher, editor and writer. He has authored books on diverse topics including natural history, earth sciences and railways, as well as hundreds of articles for encyclopedias and partworks, covering everything from elephants to abstract art. He and his family currently live near Auckland, New Zealand, where he volunteers for the local Coastguard unit and dabbles in art and craft.

The Artist

Born in Medellín, Colombia, Santiago Calle is an illustrator and animator trained at Edinburgh College of Art in the UK. He began his career as a teacher, which led him to deepen his studies in sequential art. Santiago founded his art studio Liberum Donum in Bogotá in 2006, partnering with his brother Juan. Since then, they have dedicated themselves to producing concept art, illustration, comic strip art and animation.

Picture Credits (abbreviations: t = top; b = bottom; c = centre; l = left; r = right)
© www.shutterstock.com:

1 c, 2 cl, 4 c, 6 tc, 6 br, 7 tc, 7 bl, 9 c, 11 c, 13 c, 15 c, 19 c, 21 c, 23 c, 25 c, 27 c, 28 tr, 28 bl, 29 tr, 29 cl, 29 br, 31 cr.

11 c By Dinobass (own work) [CC BY-SA 4.0 (http://creativecommons.org/licenses/by-sa/4.0)] via Wikimedia Commons

13 c Joseph Berger/© Bugwood.org/CC-BY-3.0-US

索引

作者简介

马特·特纳出生于英国，20世纪80年代毕业于拉夫伯勒大学艺术学院，毕业后一直担任图片研究员、编辑和作者。他的书题材广泛，涉及博物学、地球科学和铁路等，并为百科全书和分册出版的丛书写过数百篇文章，从大象到抽象艺术无所不包。他现在和家人住在新西兰奥克兰附近，他还是当地海岸警卫队的志愿者，平时也涉猎工艺品的制作。

绘者简介

圣地亚哥·卡列出生于哥伦比亚的麦德林，是一位插画师和动画师，曾在英国爱丁堡艺术学院接受过培训。他的第一份职业是教师，教学促使他在连续性艺术领域继续深造。2006年，他和兄弟胡安在波哥大合伙创立了一家名为“自由德南”的艺术工作室，自此两人便致力于概念艺术、插画、连环画和动画的创作。

术语表

变态发育	蝴蝶等动物在个体发育中，经历卵、幼虫（不完全变态发育的昆虫幼虫称为“若虫”）、蛹、成虫各生命阶段。不完全变态发育的昆虫跳过蛹的发育阶段，直接从若虫成长为成虫。
捕食者	捕食其他动物的动物。大约三分之一的昆虫都是捕食者。
产卵器	雌虫腹部用以产卵的中空构造，外形通常如同一根尖刺。
触角	昆虫头部细长可移动的感觉器官（通常为两根），具有触觉、味觉和嗅觉功能。
毒液	注入猎物或天敌体内的有毒液体，比如蚂蚁螫刺内的毒液。
反刍	消化后的食物从胃里返回嘴里。有些种类的蚂蚁经常用反刍出来的食物彼此喂食。
腹部	昆虫身体分为头、胸、腹三部分，腹部是最后面的一部分。
级	蚂蚁、蜜蜂、胡蜂、白蚁等社会性昆虫分化为不同的级，比如工蚁、兵蚁，各自在群体中履行不同的职能。
脊椎动物	有脊椎的动物，例如哺乳动物、鸟类、鱼类、爬行动物和两栖动物。
茧	完全变态发育的昆虫幼虫在蛹的外围做的壳。在茧里，幼虫逐渐变为成虫。
入侵性捕食者	被人类有意或无意地从原生地引入到新地域的捕食性动物，经常会打破当地的生态系统。
食腐动物	以落果、腐肉、垃圾等任何能找到的东西为食的动物。
蜕皮	对于昆虫而言，蜕皮就是脱去旧的外骨骼，以适应生长的需求。新长出的外骨骼最初是柔软的，在完成蜕皮时变硬。
吻突	昆虫（比如甲虫）的口鼻部或虫喙。
腺体	能产生化学物质，并分泌至体内或体外的器官。
胸部	昆虫身体分为头、胸、腹三部分，胸部是中间的部分。

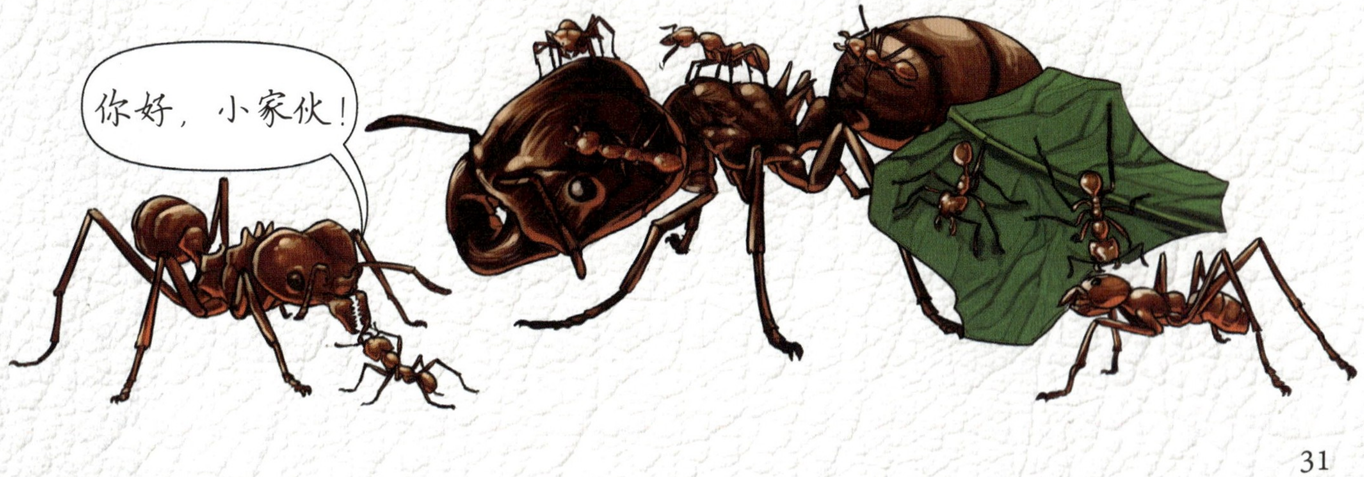

昆虫小常识

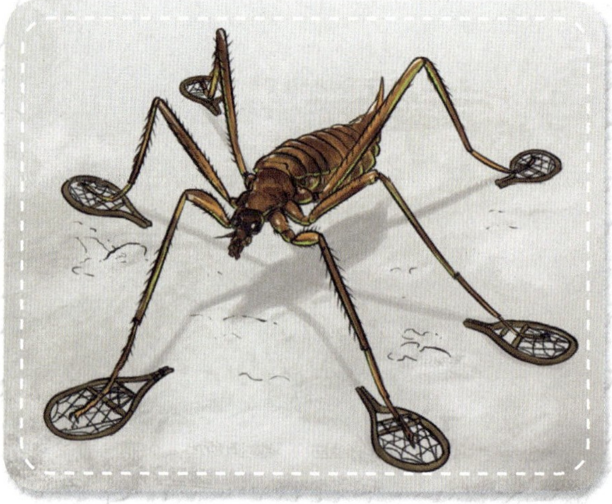

你怎么称呼没有翅膀的蝇虫？

走蝇！实际上，没长翅膀的蝇虫比比皆是，无翅雪大蚊属的雪蝇便是其中之一。在冬天，它们能在雪地上行走，可谓名副其实。雪蝇的血淋巴含有一种天然的防冻剂。蝠蝇也没有翅膀，它们寄生在蝙蝠的皮毛里，只能靠搭乘蝙蝠的便车“飞行”。

昆虫毒素可以拯救人类的生命吗？

科学家正在对多种昆虫毒素进行研究，用以研发抗癌药物。这些毒素包括第25页提到的青腰虫素，以及在花萤（又称“士兵甲虫”，译注）体内发现的罕见脂肪酸DHMA。此外，巴西胡蜂毒液中的毒素MP1似乎可以杀死癌细胞，同时让正常细胞免受伤害。

世界上有多少只蚂蚁？

无论处于哪个时间点，地球上的蚂蚁总数都在10^{15}（即1,000,000,000,000,000）只左右。

山羊能拯救昆虫吗？

老鼠、白鼬等入侵性捕食者消灭了大量的新西兰巨沙螽，人们曾一度认为这种昆虫已经在新西兰大陆上灭绝了。1962年，在新西兰北岛马霍埃努伊附近多刺的荆豆丛中，人们再次发现了巨沙螽。由于人们在这一带种植荆豆作为山羊的牧草，多刺且密集的荆豆让老鼠等捕食者难以进入，成为保护巨沙螽的天然屏障。50多年过去了，巨沙螽还在那里安然无恙地生存着。

超级蚁群

在欧洲南部，人们发现了迄今为止最大的蚁群。这个超级蚁群由30个相连的小蚁群组成，横跨6,000千米，其中的工蚁不计其数。阿根廷蚁分布广泛，从加利福尼亚州到欧洲一直覆盖到日本，也形成了一个不可思议的巨型群落。来自不同蚁群的蚂蚁通常会相互攻击，但是同属一个超级蚁群的不同小群落的蚂蚁会友好相待，这证明了它们的亲属关系。

隐翅虫大家族

要论家族规模，在所有的甲虫种群中，隐翅虫科独占鳌头。光是已命名的物种就有62,000多种，还有更多的新物种等待被发现。实际上，隐翅虫科的物种数比所有脊椎动物的物种数加在一起还要多！

高层建筑

相对于昆虫的体长而言，昆虫界的最高建筑当属非洲白蚁丘。这些白蚁丘有的高达9米，以人类的视角来看，相当于一座1,600多米高的摩天大楼。除此之外，有的白蚁巢穴还可以深入地下60多米！

破纪录的昆虫

昆虫虽然个头小，却创下了一些令人瞩目的纪录。这两页内容并不会告诉你昆虫世界里谁最大、谁最致命、谁速度最快，但你可以读到与前面内容相关的昆虫的“壮举”。有的昆虫在你家的房子或院子里就有，你可以仔细研究——不过，但愿你家的墙壁里没有藏白蚁。

夜光

发出最亮生物光的昆虫是牙买加叩头虫。发光部位是位于胸部上方的两盏“头灯”，以及腹部下方的发光区。牙买加叩头虫发出的光特别亮，以至于当地人过去常常将它们收集在葫芦里，挂在天花板上给屋子照明。

爱“换装”的虫子

蜕皮次数最多的昆虫是衣鱼。衣鱼这种古老的生物，自从恐龙时代以来几乎没什么变化。它们闯进我们的家里，啃食图书和衣服。衣鱼在一生中可能会经历60次蜕皮，甚至更多次数。准确地说，它们蜕去的是外骨骼。

白蚁

白蚁是社会性昆虫，身材虽小，影响力却很大。尽管有些种类的白蚁会毁坏房屋和庄稼，但也有许多白蚁是农民的朋友，能改善蚁群附近的土质。庞大的白蚁蚁群有着数千只甚至是数百万只工蚁和兵蚁，它们都无条件地为蚁后工作（见下图）。这是多么厉害的蚁后啊！身体长达10厘米，可能会活到50岁，这在昆虫世界里也算是创纪录了。

白蚁

体长：工蚁4—15毫米，蚁后可达100毫米及以上
寿命：工蚁和兵蚁1—2年，蚁后长达50年

“迷你牛”

白蚁的肠道分为多个室，可以分解植物中难以消化的纤维素，这一点和牛有些类似，所以有人把白蚁称为“迷你牛”。

食物链

相比其他种类的昆虫，有翅膀的白蚁为热带雨林中的哺乳动物、鸟类和两栖动物提供了更多的食物。

建筑大师

欢迎来到白蚁王国！蚁后由蚁王呵护着，寿命较长，一生致力于产卵。工蚁和兵蚁没有视力，也不能繁育后代。蚁后还时不时会产下一些长有翅膀的繁殖蚁，它们能够繁育后代，创立新的蚁群。在上图中，你能分清谁是谁吗？

白蚁既是建筑师，也是干农活的好手！有些白蚁能够建造巨大的泥塔，内有地下室、空气调节管道和菌类农场，可谓一应俱全。

兵蚁负责保卫蚁群。有的兵蚁依靠发达的颚部御敌。而上图左侧这只长鼻兵蚁（属于热带象白蚁，译注）长有一个形似喷嘴的头部，可以向蚂蚁和其他天敌喷射有毒的黏性物质。

干木白蚁是一种害虫，生活在墙壁中。它们啃食木材，每年在全球范围内造成的损失高达数十亿美元。

甲虫

甲虫是地球上种类最多的动物。很多甲虫成了鸟类、蜥蜴、哺乳动物和蜘蛛的腹中餐。为了逃避被捕食的噩运，它们进化出了一系列高超的防御本领，包括喷射致命的毒性物质、制造爆炸、分泌清洗液，以及喷气推进。甚至连花园里最不起眼的瓢虫，也能从身体里排出一些恶心的东西来自卫！

隐翅虫

体长：可达40毫米

寿命：成虫20—60天

进化

在2.7亿年的生存进化史中，甲虫的各种化学防御系统经历了30多次升级改造。

毒素

某些隐翅虫的血淋巴里含有青腰虫素，这是一种极其致命的毒素，能让前来捕食的蜘蛛望而却步。

高超的防御术

气步甲（俗称“放屁虫”，译注）可以将腹中的化学物质混合成高温毒性物质，向试图靠近它的捕食者喷射，致其死伤。

瓢虫的腿部关节会渗出血淋巴（即昆虫血），这种物质不仅味道糟糕，而且一遇到空气就会变硬，粘住攻击者的颚部。

上图左侧是棕榈龟甲虫的幼虫，它将自己的便便弄成长条，然后做成“帐篷”藏进去。位于上图中间的是棕榈龟甲虫的成虫，它的脚很大，每只脚上长有约10,000根刚毛，从脚上的腺体中还会分泌出一种黏糊糊的油脂。当受到威胁时，它会把脚黏附在树叶或其他与脚接触的平面上，安如磐石，谁都别想把它挪开，如上图右侧所示。

隐翅虫科的普通大眼隐翅虫在水面上紧急降落时，会从特殊的腺体中释放出气体，以达到喷气式推进的作用，利用“放屁的力量”滑到安全的地点。

食腐甲虫以及它们的幼虫以腐烂的动物尸体为食，它们的生活环境实在是太脏啦！好在它们会释放氨，让有害的细菌远离自己。

发光虫和萤火虫

发光虫和萤火虫如同被赋予了生命的手电筒，能自体发光。它们共有2,000多种，有些种类发光是告诫捕食者自己属于黑暗料理，会坏了食客的胃口；还有些是为了吸引配偶或者引诱猎物。这种天然光源属于生物光，是由昆虫腹部的化学反应产生的，和荧光棒的工作原理类似，二者的区别在于发光虫可以持续发光。

欧洲萤

体长：幼虫3—25毫米，
成虫15—30毫米
寿命：幼虫2—2.5年，
成虫10—16天

零损耗冷光

发光虫发出的生物光是冷光，几乎100%的能量都转化为了光能，丝毫不会以热量的形式被浪费掉。

灯光秀

有时一个地区所有的萤火虫会开启相同的发光模式，齐亮齐灭，形成超大规模的同步灯光秀。除了萤火虫，没人知道这是在搞什么名堂。

黑暗中的光亮

罗孚萤属雄萤会发出特殊模式的闪光信号，来吸引同种雌萤。雌萤会发出回应信号，意思是“到这里来”，将雄萤引导到自己身边。

有时，光萤属雌萤会耍手段欺骗罗孚萤属雄萤。它们通过闪光发出正确的回应，然后在会面地点吃掉雄萤。

在新西兰有种发光虫，它们是穴居的蕈蚊的幼虫。这些幼虫住在黏液“睡袋”里，将自己悬挂在洞穴顶部，吐出长达50厘米、挂满黏液珠子的丝线，然后“打开”尾部生物发光器的开关……

任何飞虫一旦不慎闯入悬挂的黏液丝线，都将立刻被黏液粘住；接下来，发光的蕈蚊幼虫用丝线将飞虫绕到自己跟前，大快朵颐。

在美国，一些甲虫的幼虫会通过发光的方式来引诱马陆之类的猎物上钩。因为它们看起来像是在黑夜中亮着灯的火车，所以被称为“铁道虫”。

螳螂

螳螂是伏击高手，善于伪装捕食。当它们栖息在植物上时，能和周围环境融为一体。螳螂用大大的复眼观察周围有无风吹草动，一旦有其他昆虫靠近，便会以迅雷不及掩耳之势，伸出长长的、带有尖刺的前腿抓住它。螳螂的种类很多，伪装形式可谓五花八门——有的伪装成五颜六色的花朵，还有的伪装成枯枝败叶。

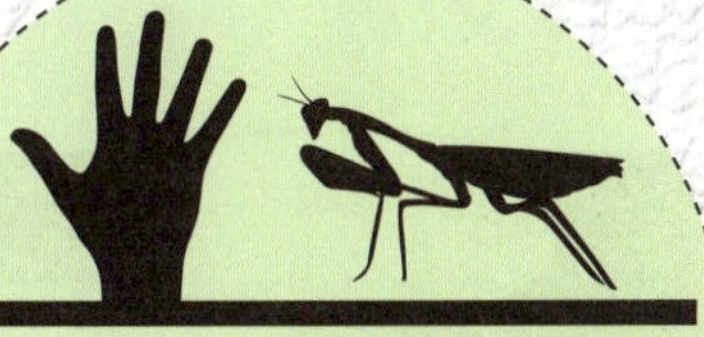

小提琴螳螂

体长：雌虫10厘米，
雄虫8—9厘米
寿命：1年，有时长达2年

我有刺，别惹我！

小提琴螳螂胸部超长，就像小提琴细长的琴颈部分。它的外形还酷似蔷薇的枯枝，这也正是它喜欢藏身的地方。

玩失踪

螳螂的若虫在蜕皮时可能会改变体色，比如从褐色变为绿色，为的是更好地融入周边的环境。

酷酷的伪装

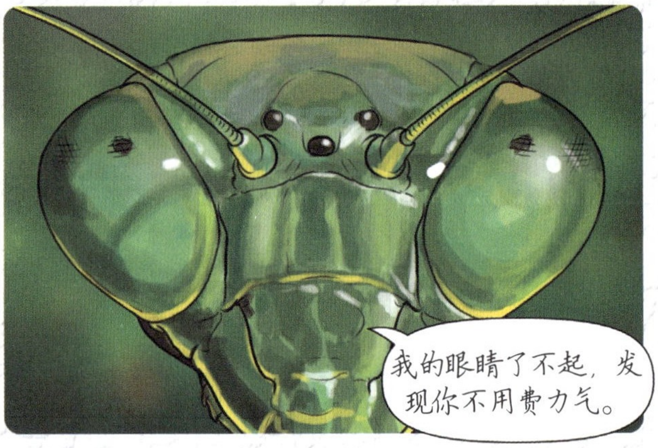

在所有昆虫中，唯有螳螂能将头向左右两侧各转180度。所以，别看螳螂栖息在某处一动不动，其实它们正在密切监视猎物呢。

螳螂的两只眼睛上各“浮”着一个黑点，这并不是真正的瞳孔，而是用来吸收光的部位，眼睛的其他部分则用来反射光。

和叶䗛类似，螳螂走起路来也左摇右摆的。它们这样做是为了假装成在微风中摇动的小细枝，增强伪装效果。而上图左侧这只花螳螂则是在摆出一种“吓唬姿态”，炫耀它靓丽的颜色。它这样做有时是为了吓走捕食者。

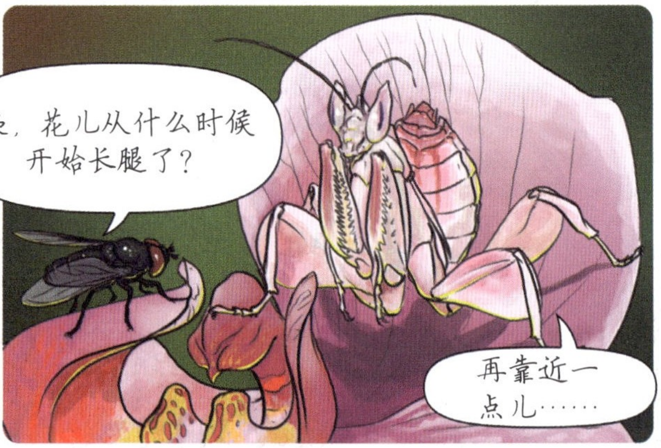

一些昆虫的伪装达到了极致。兰花螳螂就是个例子。它们藏在花丛中，想把它们分辨出来可没那么容易！

螳螂的卵囊看起来就像挤出来的一截牙膏，经常挂在花园的篱笆上。新孵化出的若虫就像是迷你版的螳螂。

叶䗛

叶䗛被称为“会行走的叶子”。它们将伪装的本领发挥得淋漓极致，经过漫长的进化，其外形与真正的叶子别无二致，不仅有“叶脉”，甚至还有卷曲的褐色“叶缘”。它们与竹节虫是近亲，同属于竹节虫目。和某些种类的胡蜂类似，雌性叶䗛能够在不与雄虫交配的情况下产卵，孵化出更多的雌虫！

巨叶䗛

体长：可达11厘米

寿命：5—7个月

稀有的雄虫

尽管叶䗛的雄性个体是存在的，但其罕见程度让人难以置信。直到1994年，人们才发现了第一只巨叶䗛的雄虫。

同类相食

把竹节虫当宠物养是个不错的主意。但是如果没有叶子吃，它们会吃叶䗛……所以，不要把它们放在一个笼子里养！

酷似叶子的昆虫

1519年至1522年，安东尼奥·皮加费塔在菲律宾探险时发现了叶䗛。他认为它们是“会行走的叶子”，以空气为食。

叶䗛若虫在发育过程中会蜕好几次皮。如果它们掉了一条腿或一根触角，都可以重新长出来——伴随着每次蜕皮，腿都会变长。

竹节虫目昆虫的夜视能力让人吃惊。每次蜕皮，它们的眼睛都会变得更大、更敏感。所以，若虫习惯白天出来活动，而成虫就基本上算是个夜猫子了。

某些叶䗛会从颈部的腺体中喷射出一种防御性的液体，以此来阻止捕食者的骚扰。如果这种液体进了你眼睛里，你会感到非常疼痛。

在叶䗛的近亲中有一种双纹竹节虫，它们又被称为“恶魔骑士”，是美国的一个物种。在夏秋时节交配期，雌虫和雄虫会抱在一起。不管去哪儿，两口子都黏在一起，直到其中一方死去，掉落在地上，它们才会分开。

巨沙螽

什么昆虫不会飞，吃素，面孔像日本武士，颚部强健，腿部长有尖刺，体重赛过麻雀？对了，这就是巨沙螽！

巨沙螽分为11个种类，全部生活在新西兰。那里的老鼠、白鼬和其他入侵性捕食者已经将大陆上的巨沙螽消灭殆尽，只在海岛和自然保护区留下了仅有的幸存者。还有一些种类的沙螽体形较小，它们生活在洞穴、森林和高山之中。

巨沙螽

体长：可达10厘米

体重：可达70克

寿命：2年左右

小岛上的巨人

岛屿巨型化指在岛屿上孤立进化的动物或植物随着时间的推移变得巨大无比。巨沙螽就是一个岛屿巨型化的例子。

丑陋的沙螽

新西兰毛利人将“沙螽”称为“威塔庞加”，借用的是丑物之神的名字，意思是“幽灵”或“鬼”。

令人咂舌的庞然大物

巨沙螽的体重通常能达到35克，比老鼠还要重。被正式记录的史上最重的巨沙螽是一只怀孕的雌性巨沙螽，它的体重达到了70克。

雌性沙螽和它的近亲蟋蟀一样，也长着一个尖尖的产卵器。雌性沙螽产卵时，会将卵推入潮湿的土壤里。

山沙螽生活在雪线（终年积雪区和融雪区的分界线，译注）之上，即使冻僵了也不会死！当天气回暖，它“解冻”以后还能活下去。

1970年，人们发现了长牙沙螽。雄虫用长牙打斗，争夺雌虫，还能摩擦两根长牙，制造出一种刺耳的噪音。

洞穴沙螽算不上什么庞然大物，但能算是个“大长腿”。它们的身长仅有约3.5厘米，跳跃一次的跨度却长达21厘米。在新西兰阴冷潮湿的洞穴里，你会发现很多洞穴沙螽附着在洞穴顶部。如果你去洞穴里寻找它们，可得记住在出洞之后检查大衣领哦！这些“大长腿”可都是跳高能手！

象鼻虫

象鼻虫在甲虫中体形偏小，通常长有一个长鼻子状的口鼻部，触角上还有一个类似于肘关节的拐角，已知的种类超过60,000种。虽然象鼻虫爱刨植物，毁害庄稼，因而深受农民厌恶，但它们所具有的某些适应性特征让人叹为观止。有些种类的象鼻虫口鼻部（也叫做“吻突”）极长，雄虫用它来相互格斗，争夺雌虫。

马达加斯加长颈象鼻虫

体长：可达25毫米

寿命：1年左右

数量庞大

在所有生物中，差不多有四分之一是甲虫，而甲虫中大约有五分之一是象鼻虫。象鼻虫简直是无处不在！

貌如其名

雄性马达加斯加长颈象鼻虫的口鼻部几乎占据了整个身体的长度！交配后，口鼻部短得多的雌性象鼻虫会小心翼翼地把叶子卷起来，在每片叶子里产一枚卵。

钻孔专家和战斗勇士

上图所示是来自新西兰的雄性长颈象鼻虫，它们的模样和下一页的马达加斯加长颈象鼻虫很不同，但二者都可能以“长鼻子大战”的胜负来决定谁能与雌虫交配。新西兰长颈象鼻虫雄虫的整个身体长达9厘米，巨大的口鼻部便占去了一半。

对于栎实象甲来说，雌性个体的口鼻部更长。它用口鼻部在橡子上钻出一个孔，然后转过身来，在孔中产下一枚卵。

幼虫在橡子中发育，然后钻破外壳，从橡子里出来，落入土里化为蛹，经过一两年的时间变为成虫。

棕榈象甲在椰子树、海枣树和油棕树上一次能产下多达500枚卵。巨大的幼虫咬穿木材，毁坏农作物。

昆虫富含蛋白质。在马来西亚，人们用象鼻虫的幼虫做菜，名叫“象鼻捞”。在越南，人们将活的象鼻虫蘸鱼露吃。很美味哦！

蚁狮

蚁狮不是蚂蚁，在世界范围内已发现大约2,000个种类。虽然成虫长有翅膀，看起来就像一只瘦巴巴的蜻蜓，但幼虫则是一个凶残可怕的小怪物。有时，蚁狮也只是藏在树叶中或岩石缝里，但它名声在外的习惯是在沙土里刨坑——然后在那儿张大颚部，伺机而动，擒获猎物，并将猎物的体液吸食殆尽。

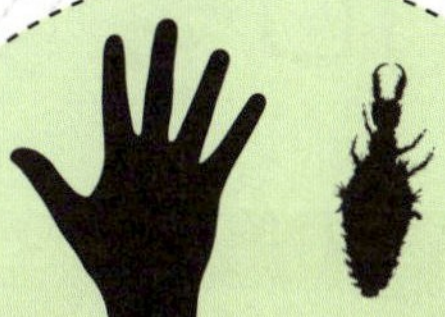

蚁狮

体长：幼虫可达15毫米，
成虫可达80毫米

寿命：幼虫3年及以上，
成虫25—45天

隐匿的杀手

“蚁狮”这个名字已经存在好几个世纪了。中世纪学者莫鲁斯在谈及蚁狮时曾经提到：“它藏身于沙土中，捕杀搬运食物的蚂蚁。”

昵称

一些美国人将蚁狮称为“涂鸦虫”。在马克·吐温的经典小说《汤姆·索亚历险记》中，汤姆和一只“涂鸦虫”说话，试图引诱它出洞。

可怕的幼虫

在这里产卵实在是完美！

这沙坑挺小巧的！

蚁狮是一种完全变态发育的昆虫，在个体发育中经历卵、幼虫、蛹、成虫四个时期，从而构成完整的生命周期。这一切，首先从蚁蛉（即蚁狮成虫）雌虫在泥土或沙子里产卵开始。

孵化出的幼虫挖出一个锥形的坑，把身体埋在坑底，只露出长长的、长满“牙齿”的颚部，然后埋伏以待。

你能来一起玩，真是太好啦！

如果猎物不慎落入，蚁狮会快速弹射沙子引起“沙崩”，使猎物措手不及，并从隐藏处猛冲出来，通过颚部的空心“牙齿”将具有消化作用的毒液注射到猎物体内，制成“靓汤”后吸食，最后只留下猎物的空壳。

你可能会觉得这挺无聊的。

我足足等了三年，就为了这个?!

幼虫在一个鹰嘴豆大小的球状茧里化蛹。蛹可能在沙子里埋上好几年。

最后，蛹羽化成又细又长的成虫，并长出一对翅膀。在接下来的几个星期，它不吃不喝，一心寻找配偶。

蚂蚁

大约1.3亿年前，蚂蚁才从古老的胡蜂进化而来，不过现在它们的足迹已经遍布世界各地了。蚂蚁是群居动物，一个蚁群的蚂蚁数量有时多达数百万只。蚁群中有明确的“级”，蚁后、工蚁、兵蚁等各司其职。蚂蚁们相互协作，一起筑巢、抚育后代、抗击敌人。为了蚁群的生存，它们愿意战斗到死。

斗牛犬蚁

体长：8—40毫米

寿命：长达2年左右

凶猛的斗士

斗牛犬蚁性情凶猛。即使身首异处，它的头仍然会试图噬咬腹部，腹部的毒刺会蜇向头部。

比比谁更重

如果你能把一架巨型天平带到亚马孙雨林，把那里所有的蚂蚁放到天平的一端，所有的陆生脊椎动物（包括哺乳动物、鸟类、爬行动物和两栖动物）放到天平的另一端，那么蚂蚁的重量将是另一端的四倍左右。

蚁心齐，泰山移

大大小小的蚂蚁分工合作。上图左：一只大切叶蚁在搬运一块碎叶子，它让几只小切叶蚁搭了个便车；作为回报，小切叶蚁帮大切叶蚁驱赶寄生蝇。上图中：一只超大巨首蚁为小巨首蚁充当“军用运输车”。上图右：一只行军蚁工蚁正在帮一只兵蚁清理颚部。

一只织叶蚁轻轻地叼住一只幼虫，用触角轻拍幼虫让它吐丝。织叶蚁用这种丝将叶子黏结起来，修建蚁穴。

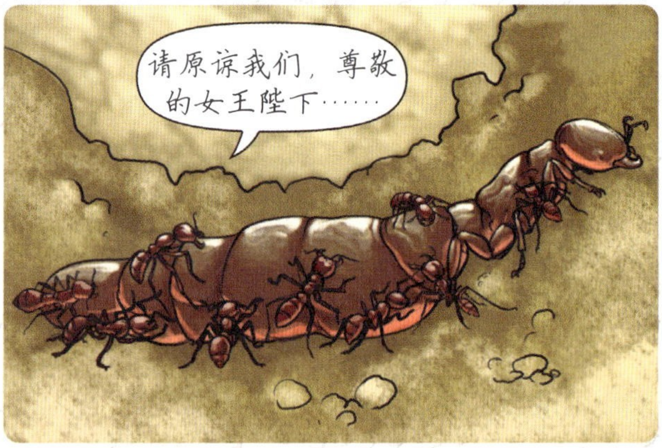

矛蚁的蚁后体长可达5厘米。因为蚁后块头太大，行动起来要靠只有它十分之一大小的小工蚁推它。

一些木蚁群中的工蚁将防卫工作做到了极致。当遭遇攻击的时候，它们会通过“自爆”的方式，向敌人喷射有毒的黏性物质。

蜜蚁的工蚁又称“贮蜜蚁”，它们挂在蚁穴顶部，从腹中反刍出液态的蜜糖和蛋白质来喂养其他蚂蚁。

蟑螂

晚上，当你打开厨房的灯，可能会在地上看到一只扁平的、褐色的昆虫迅速冲到暗处藏起来。很少有昆虫像蟑螂一样不招人待见。然而，在已知的大约4,600种蟑螂中，只有大约30种是害虫。有些种类的蟑螂为适应寒冷或干燥的栖息环境做出了特殊改变。然而，大多数蟑螂都具有共同的特征，它们“吃嘛嘛香”，待哪儿都觉得舒坦。正因为这样，在地球上的几乎每个角落都有蟑螂出没。

马达加斯加发声蟑螂

体长：可达76毫米

寿命：2—5年

抗核辐射

果蝇和蟑螂在受到核攻击后能够存活下来。它们可以抵抗大剂量的核辐射，而同样剂量的核辐射足以杀死人类。

喜好甜食

如果蟑螂尝过香草味或者薄荷味的甜食，那么今后当它察觉到空气中有这种气味时，它便会充满期待地流口水。

超级生存者

一只掉了脑袋的蟑螂可以存活好几个星期。在断下来的脑袋上，那对触角还能继续摆动好几个小时！

蟑螂最早出现在大约3.2亿年前。我们今天看到的这种蟑螂在大约2亿年前就已经出现了，它们曾经和恐龙共同行走在地球上。

蟑螂是食腐动物。当食物匮乏时，蟑螂几乎无所不吃：胶水、油脂、肥皂、墙纸胶、皮革，甚至是毛发，都能被它们填进肚子。

有些种类的蟑螂体形巨大。中美洲长翅蟑螂的翼展可达17.5厘米；澳大利亚犀牛蟑螂重达33.5克。

蟑螂是“飞毛腿”！一旦六条腿都动起来，蟑螂能以高达每秒1.5米的速度全力冲刺。不但如此，它们还灵活机敏，即使在全速前进时仍能在犄角旮旯的地方急转弯。

一些昆虫独居，但更加高级的昆虫，比如蚂蚁、蜜蜂和白蚁，会组成复杂的社会群体，共同建造巨型建筑、采集食物、抚育后代、抵御外敌。

在本书中，你将会读到现代昆虫因适应生存环境而形成的一些最令人惊叹的特征，包括它们美丽的外表、伪装的本事、设计建造的技能、感官、武器、防卫术和为生存所采用的诈术，以及它们给人类带来的福与祸。在你的家里和花园里就有一个昆虫动物园，是不是很奇妙呢？一起去看看吧！

不可思议的昆虫

我们人类可能自以为是地球的主宰者，殊不知昆虫才是真正的老大！

首先，昆虫的数量远远多于人类。据粗略估算，在地球上任何一个时刻，昆虫的数量都达到了约10^{19}（10,000,000,000,000,000,000）只，种类超过90万种。其次，昆虫远远比人类古老，其起源可以追溯到将近5亿年前，比恐龙时代还要早得多。昆虫比有花植物也要早出现大约1.2亿年。换句话说，当有花植物还在为生死存亡进行漫长的进化时，昆虫早已在地球上占据了一席之地，而不是相反的情况。

目录

在任何一个时刻，地球上都有大约 10^{19}（10,000,000,000,000,000,000）只昆虫，它们已经生存了将近五亿年，许多种类都成群生活。打开这本书，一起来走近地球上这些伟大的幸存者吧！在这本书中，你将发现：

- 地球上所有蚂蚁加起来的重量超过人类的总重量
- 螳螂酷似一朵花或是一片腐烂的树叶
- 有些白蚁蚁后一天可以产下多达 40,000 枚卵

……

还有更多关于昆虫的奇闻轶事，比如蚂蚁军团、“以假乱真”的叶䗛（它们最初被误认为是行走的叶子）、建造高塔的白蚁和闪烁的萤火虫。昆虫统治地球，难道不是吗？

番茄博士双语百科

不可思议的虫子王国

昆虫三十六计

马特·特纳（英）著
圣地亚哥·卡列（英）绘
丛岚 译

外语教学与研究出版社
FOREIGN LANGUAGE TEACHING AND RESEARCH PRESS
北京 BEIJING

Record-Breaking Bugs

Flying Insects

Written by Matt Turner
Illustrated by Santiago Calle

外语教学与研究出版社
FOREIGN LANGUAGE TEACHING AND RESEARCH PRESS
北京 BEIJING

Flying insects hold one of the great keys to survival: wings. These powerful structures turn bugs into jewels and make for amazing aerial acrobatics. Get close to discover how

- the hummingbird hawk moth dodges predators as it feeds
- honeybees dance to show where food-rich flowers are
- one emperor dragonfly camouflages itself on the wing

...

and more flying aces. You will also read the stories of master-builder wasps and bees, carnivorous robber flies, butterfly pupae disguised as leaves and the always pesky mosquito. Dazzling! It's enough to make you wish you had wings too!

Whoa!
Didn't see that coming.

CONTENTS

Flying Insects

Why did the first insects fly? You might as well ask why plants began producing flowers, about 150 million years ago. The answer in both cases is 'because it helped them survive'. And in fact, plants produce their colourful, scented flowers to attract insects and other animals to pollinate them. The first winged insects appeared on Earth around 400 million years ago, when plants began to grow taller and produce the first forests. So having wings enabled early insects to evade predators, and later to feed on pollen and nectar – just as bees, butterflies and others do today.

Wings are so important to insects that most of the major groups are named after them, using the Greek word *pteron* for 'wing'. For instance, dragonflies and damselflies, which were among the first insects to take to the air, belong to the Palaeoptera ('old wings') group. These descriptive names can help us identify insects. Beetles (Coleoptera = 'hardened wings') keep their hind wings folded away under tough, protective forewings. Mayflies (Ephemeroptera = 'short-lived, winged') have a brief adult life, often lasting just a few hours.

This book includes winged insects that are especially big or beautiful, fast or skilful, or dangerous – plus a few more that simply show their amazing diversity. And there are many, many more out there; you'll find a lot of them in your own neighbourhood.

Mega Wings

Relax, buddy; I was just after some nectar!

When a male butterfly courts a female, he hovers over her and showers her with perfume-like scents, which are called pheromones. This chemical courtship encourages her to mate with him. The male also chases away rival butterflies – and even birds!

Phew, you're stinky!
You're not so fresh, either.

Birdwing caterpillars taste bad! This is because they feed on toxic vines, and the poisons gradually build up in their body tissues, surviving right through to adulthood. They also have an osmeterium – a 'stink organ' behind the head that helps ward off intruders like this possum (above).

Just ignore me.

A birdwing pupa, tied to a twig by a fine silken halter, looks just like a curled-up leaf. This brilliant disguise helps hide it from predators.

You taste fine to me!

About the only predator that will tackle a birdwing is the *Nephila* orb weaver, a very large spider that doesn't mind the bad taste of the butterfly's toxins.

BIRDWING BUTTERFLIES

These beautiful butterflies belong to the swallowtail group. With their long wings and strong flight, they have been likened to birds – hence the name. They are also the biggest of all butterflies. Queen Alexandra's birdwing, found in Papua New Guinea, has the largest wingspan: nearly 254 mm. Thanks to the toxins in their diet, birdwings are rarely troubled by predators.

CAIRNS BIRDWING

Lifespan: up to 3 months
Size: body length 70 mm, wingspan 125–150 mm

VIVID MALES

Rajah Brooke's birdwing is the national butterfly of Malaysia. Like other birdwings, the green-and-black male is more vividly marked than the female.

COLLECTING

There are strict laws on collecting birdwings, which are rare in the wild. Most species can be raised in captivity, though.

Humming and Hovering

The hummingbird beats its wings up to 70–80 times a second while hovering to feed from flowers, and gives its name to the hummingbird hawk moth.

The sphingids are big, strong-flying moths. They include the white-lined sphinx of North America (right), which also hovers like a hummingbird.

In 1862, Charles Darwin studied a huge orchid from Madagascar, and wondered what insect pollinated it. The moth (with mega proboscis) was found in 1903.

The hummingbird hawk moth is an impressive flier, able to 'side-slip' while hovering in order to dodge predators, such as this praying mantis.

Some flies have an extra-long proboscis, too. When similar features appear separately in unrelated animal groups, it's called convergent evolution.

HUMMINGBIRD HAWK MOTHS

If you spot a bulky moth hovering in mid-air next to flowers, collecting nectar with its very long, thread-like proboscis, it's probably one of several moths in the Sphingidae family that have evolved superb flight skills. They include the snowberry clearwing or 'flying lobster' of North America, and the hummingbird hawk moth of Europe and Asia. Listen closely and you may hear the wings gently hum.

HUMMINGBIRD HAWK MOTH
Lifespan: 7 months
(including hibernation)
Size: wingspan 40–45 mm

HIBERNATION
In parts of its range, the hummingbird hawk moth hibernates, in a cosy nook in a tree or rock, from October until April.

REPRODUCTION
The female hummingbird hawk moth produces up to four batches of eggs in the year, laying them on bedstraw *Galium*, the caterpillars' food plant.

Life is Sweet

The queen (centre) founds the hive, and lays up to 1,500 eggs a day. Fertile eggs hatch into workers (right), which forage for food, clean the hive and nurse the young (far right). They also sting! Unfertilised eggs hatch into drones (left) – males whose sole job is to fertilise the young queens that will found new colonies.

Domesticated honeybees live in artificial hives, but their wild ancestors would have nested in tree holes and similar crannies.

In doing the figure-eight 'waggle dance' for others in the hive, a bee shows them the direction from the sun to find a source of nectar, pollen or water.

Giant Asian cousins of the honeybee have a painful sting, but farmers welcome them because they pollinate crops such as cotton, mango, coconut, coffee and pepper.

Bees are vital pollinators, but are threatened by overuse of pesticides, and by Varroa, a parasitic mite that has recently spread fast.

HONEYBEES

As well as making yummy honey, bees pollinate one-third of all the crops we grow. These highly social insects live in a hive containing 50,000 or more workers and drones, ruled by a queen. Using wax squeezed from their abdomens, they build combs made up of thousands of cells, where they store their collected pollen, honey and other food, and raise larvae.

WESTERN HONEYBEE

Lifespan: worker 1–11 months, queen 2–5 years

Size: body length 10–20 mm

DANCERS

After returning from foraging, bees perform 'waggle dances' and other moves, which tell members of the hive where to find food-rich flowers.

THERMOSTATS

By vibrating their flight muscles, bees can control the temperature inside the hive, keeping it constant whatever the weather.

FUZZY AND BUZZY

Think I'm gonna sneeze...

The bumblebee is an important pollinator of early-flowering crops, such as apple. It gathers pollen in the fringed pollen baskets on its hind legs.

Do not try this at home with earwax.

A single queen founds a new nest each spring, after overwintering. She builds the starter cells by wax, squeezed from her abdominal joints.

HOME
Sweet
HOME

Compared to the orderly structure of a honeybee comb, the bumblebee nest is an untidy clump of cells built in a tangle of foliage or a hollow. The cells serve as honeypots, pollen stores and brood chambers for larvae. Workers keep the nest tidy, removing any dead bees.

Will you 'bee' mine?

In the fall, gynes (young queens) mate with drones (males). The drones later die, and the gynes feed heavily, fattening up for their winter sleep.

'Bee' off with you!

A cuckoo bumblebee is a female that invades a nest, ejects the resident queen (above), then lays her own eggs and gets the resident workers to care for them.

BUMBLEBEES

Fat and furry, the bumblebee is a welcome garden visitor because it pollinates wild plants as well as crops. Bumblebees are social, like honeybees, but form much smaller colonies, usually up to around 400 individuals. Unlike honeybees, bumblebees can fly in chilly weather, so they mostly live in cooler parts of the world. But this means their colonies usually die off in autumn, and young queens need to found new ones in spring.

BUFF-TAILED BUMBLEBEE

Lifespan: worker 4–7 weeks, queen 1 year
Size: worker body length 11–17 mm, queen 20–22 mm

WEIGHT

Aerodynamic 'experts' once calculated that bumblebees were too heavy to fly! But they had forgotten that air behaves differently around small bees than it does around big airplanes.

DUMBLEDORE

One old name for the bumblebee is 'dumbledore'. Harry Potter author J.K. Rowling used it for her Hogwarts headmaster, whom she imagined humming tunes to himself.

Expert Engineers

You can identify a paper wasp by its long hind legs, which trail during flight. By contrast, a common yellowjacket keeps its legs 'tidy'.

Paper wasps often collect timber from old fence rails. Listen carefully, and you'll hear a faint crunching as they bite it off.

Breeding males and females gather at a competition known as a lek, where males jostle for a high-up perch, or fight. Looks are important, too: females with the greatest number of black facial marks, and males with the neatest abdominal markings, are considered the most desirable mates.

Paper wasps prey on lots of different insects. In summer you may see them take caterpillars, as a source of protein for their own growing larvae.

The parasitic paper wasp *Polistes sulcifer* cannot build her own nest, so she takes over others and forces the workers feed her larvae – like the cuckoo bumblebee on page 14.

PAPER WASPS

First seen in the US around 1980, European paper wasps have since settled across the country – driving out native paper wasps, and collecting insect prey to feed their young. The nest is a marvel of engineering: a bulbous cluster of open cells, usually suspended from a narrow waist or pedicle. Paper wasps often roam the garden in packs; though less fierce than yellowjackets, they will sting if you get too close.

EUROPEAN PAPER WASP

Lifespan: queen up to several years, worker and drone 11 months
Size: body length 8.5–13 mm

NESTS

Paper wasps are named after their 'paper' nest-building material, which is actually chewed-up timber and spit.

WATERPROOF

Paper wasp spit is amazingly waterproof, and scientists have copied its formula to create a special coating for unmanned aircraft.

FLYING ACES

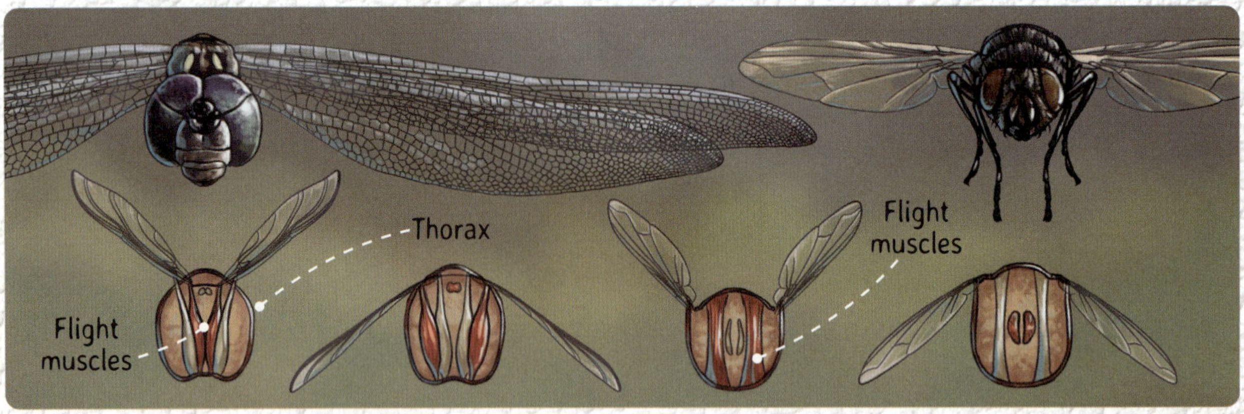

The dragonfly's wing arrangement (top left) is primitive but powerful. At its root end, each of the four wings is attached to a flight muscle, for direct and independent control. A more modern insect, such as a housefly (top right), uses indirect flight: it deforms its thorax to flex the flight muscles, as these cross-sections show.

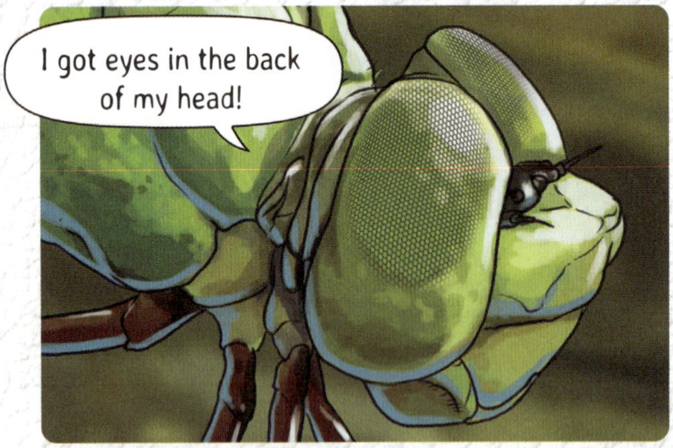

The huge compound eyes typically meet at the top, for 'wrap-around' eyesight. In some dragonflies the eyes are wide-set for better binocular vision.

The predatory larva lives up to two years on the pond bed. It shoots out its mask (a hinged lower lip) in less than 25 milliseconds to snatch prey.

Males are territorial, defending a patch of pond and chasing rivals away in aerial combat. The Australian emperor *Hemianax papuensis* uses motion camouflage, a special trick where he takes a flight path that makes him seem motionless to his rival, who can no longer pick him out from the background landscape.

DRAGONFLIES

Swooping and diving after their aerial insect prey, dragonflies are the supreme flying aces – thanks in part to their all-round eyesight, and also to the way the big flight muscles in their thorax attach directly to the base of each wing. Europe's biggest dragonfly, the emperor, can hunt non-stop for hours over lakes and rivers, especially when there's plenty of sunshine to warm its muscles.

EMPEROR DRAGONFLY

Lifespan: adult usually 4 weeks, but up to 8.5 weeks
Size: body length 66–84 mm, wingspan average 106 mm

WINGSPAN

Along with damselflies, dragonflies form the order Odonata. Largest of them all is the giant helicopter damselfly of Central America, with a 191-mm wingspan.

DIRECTIONS

A dragonfly can fly in any direction, and even upside down, when chasing prey or rivals. But due to the set-up of its forelegs, it cannot walk.

LIGHTNING FAST

The halteres, which vibrate during flight, detect any changes in the fly's pitch, roll and yaw movements (up-and-down and side-to-side). They send signals to the nerve tissues in the thorax, enabling the flight muscles to constantly stabilise the fly's body. (If you want to see halteres, crane flies have a big pair.)

When you try to swat a fly, it calculates the angle of the approaching threat and within 0.01 seconds it has adjusted its legs to spring off in a safe direction.

Blow-flies are often the first flies to lay eggs on a carcass, because their maggots eat rotten meat. Police experts look for blow-flies on a corpse to calculate when it died.

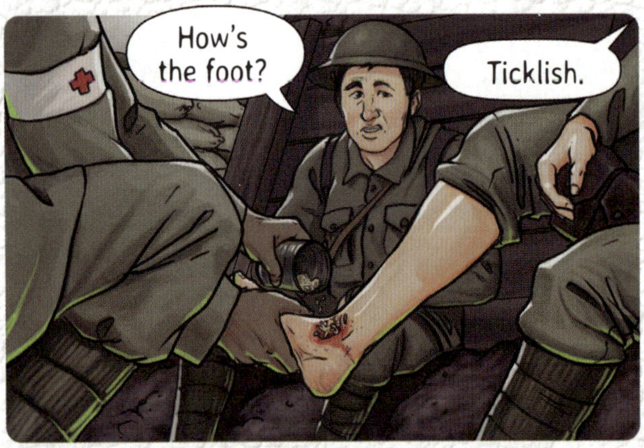

Because maggots only eat dead flesh and not healthy tissue, they have been used to clean up festering wounds – on soldiers in wartime, for example.

Scientists are researching natural chemicals in blow-fly maggots. This may lead to new drugs to fight resistant infections, or even cancers.

BLOW-FLIES

So how does a fly fly? In the true flies (order Diptera), the secret lies partly in the halteres. You can read opposite to learn how these clever gizmos work. Blow-flies, meanwhile, are those metallic-looking pests that perch on fresh roadkill, or on a juicy steak in the kitchen, laying their eggs and spreading disease. Their larvae – the classic horror-movie maggots – have some uses in medicine that may surprise you.

BLUEBOTTLE

Lifespan: adult usually up to 1 month; longer if it survives hibernation

Size: body length 10–14 mm

EVOLUTION

A fly flies only with its forewings. Long ago the hindwings evolved into halteres, a pair of knob-tipped stalks attached to the thorax.

REFLEXES

Why are flies so hard to swat? Partly because of their all-round vision; also because they think much quicker than we move!

SWOOPING HUNTERS

Robber flies usually hunt on sunny days. They tend to perch in wait on a plant, then zoom off and attack in flight, using their excellent vision and superb flying skills to zero in on the flight path of their prey. Spines on their strong legs help them grab the victim – and they always choose prey small enough to grab quite easily.

Many robber flies have evolved a warning colouring and appearance that helps them evade predators. For example, this fly (left) looks just like a bee (right).

The mystax is a 'moustache' of bristles above the robber fly's mouthparts that protects its eyes from struggling prey.

The fly injects toxic saliva to paralyse the prey and soften its guts, which it then sucks up. The strong proboscis of a big robber fly can punch through a beetle's wingcases.

In one group of robber flies, the male has extraordinary hind-leg plumes, perhaps to impress a female during his courtship 'dance'.

ROBBER FLIES

Not all flies feed on sugar, poo and rotting meat. The Asilidae are a worldwide family of more than 7,000 species of robber fly: fast, powerful, day-flying predators equipped with stout legs, specialised in intercepting insect prey on the wing and stabbing it with their needle-sharp proboscis. It injects saliva that paralyses the victim and dissolves its guts. Not for nothing are they also called assassin flies.

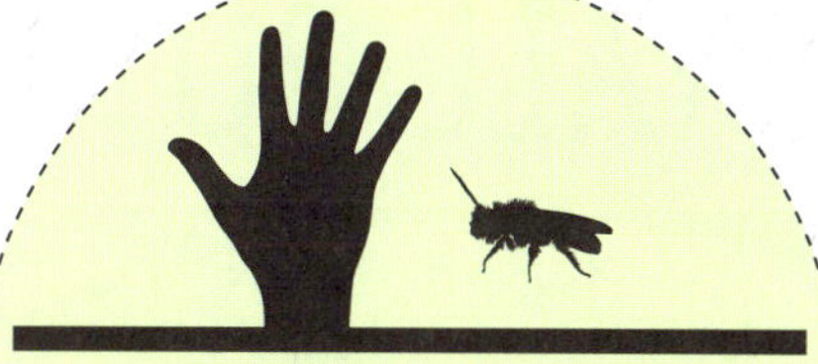

ROBBER FLIES
Lifespan: 1 year, maybe 2
Size: body length 3–80 mm

PEST CONTROL
Preying on ants, beetles, wasps, flies, grasshoppers, bugs – anything, in fact, that moves – robber flies are useful controllers of insect pests.

BIRD FOOD
Big robber flies like the bee panther *Promachus rufipes* or beelzebub bee-eater *Mallophora leschenaulti* have been known to attack hummingbirds.

Frail but Deadly

A female mosquito can locate a host by sight, but is also attracted by smells – such as sweat and exhaled breath – from as much as 50 m away.

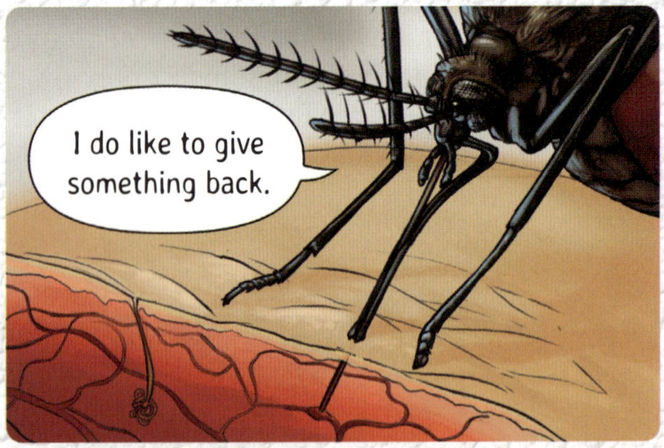

A feeding female injects saliva that stops a host's blood clotting and keeps her proboscis from clogging. It's when the saliva is infected that she can spread disease.

A male's bushy antennae can hear the special whine of an unmated female's wingbeat. If they are 'pairable', the two mossies will 'harmonise', like two people who begin to hum the same tune, and then track each other down. It's a bit like having a built-in 'find my mate' app!

Scientists took many years to realise mosquitoes spread disease. Medical pioneer Patrick Manson (1844–1922) tested malarial mosquitoes on his gardener.

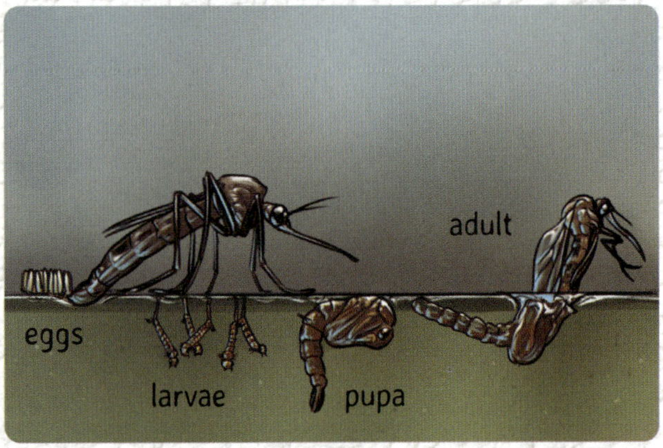

Most mosquito larvae develop in water, breathing at the surface and feeding on tiny particles. They also pupate and emerge at the surface – then fly off to find a mate.

MOSQUITOES

These flimsy little flies are parasites, the female sucking a host's blood to nourish the eggs in her abdomen. Some species also pass on diseases that kill at least two million people every year, making the mosquito the world's deadliest animal. And mosquitoes are terribly good at what they do: more sensitive to sound than any other insect, they create that awful whining noise just to find a mate.

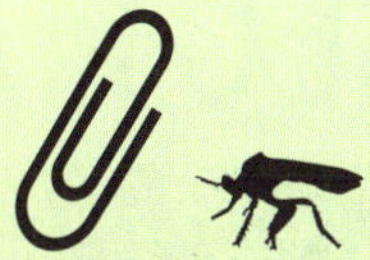

MOSQUITO

Lifespan: a few days to 1 month or more

Size: body length 2–19 mm, average 3–6 mm

DISEASES

Mosquito-borne diseases include malaria, dengue fever, hepatitis, West Nile fever, yellow fever, filariasis and Zika virus.

HIS 'N' HERS

Female mossies take blood from mammals, birds, fish and reptiles, and even from other invertebrates. Males just feed on plant juices.

INSECT ATHLETES

BAH-D-O-I-N-N-G-G-G-G-G

To jump, a grasshopper contracts two muscles in each hind-leg femur (the thick upper section) so that they pull against each other, storing lots of strain energy. Then, by suddenly relaxing one femur muscle, the energy is released and the tibia (thin lower section) can kick down to fling the insect skywards.

To attract a mate, a grasshopper makes a chirping sound, called stridulation, by moving its hind legs, rubbing them against notches on its wing veins.

Whoa! Didn't see that coming.

Some grasshoppers are brightly coloured to warn birds they are bad to eat. Others (above) will suddenly open patterned wings to surprise predators and buy time to escape.

I have a cunning plan...

Environmental changes can trigger grasshoppers to gather and migrate in huge swarms of nymphs (young) and adults, devastating crops and even causing local famine.

Revenge is a dish served crispy!

The colourful grasshoppers are usually toxic, but the edible species are full of protein. It is said that, worldwide, around 1,400 insect species are included in the human diet.

SHORT-HORNED GRASSHOPPERS

Short-horned grasshoppers don't really have horns; they have short antennae, unlike their cousins the katydids or bush crickets (with long antennae). The 10,000-odd species include the infamous locusts, which may form swarms many millions strong and strip a crop field bare in a few hours. While they can fly perfectly well, grasshoppers are far better known for their athletic leaping skills.

SHORT-HORNED GRASSHOPPER

Lifespan: adult 7–8 weeks
Size: body length up to 150 mm, average 38–50 mm

GREEDY

It's estimated that, in a single day, a one-ton locust swarm can eat as much cereal as 2,500 people. In the Bible, a plague of locusts devoured Egypt's crops.

CATAPULTS

A grasshopper's hind legs are basically a pair of catapults, designed to produce leaping energy that is both rapid and powerful.

OTHER FLYING INSECTS

There are so many beautiful and fascinating insects in the world that one book isn't nearly enough to describe them – so here are just a few more.

Diving beetle: This big, powerful predator lives in fresh water, where it snatches other insects, small fish and tadpoles in its sharp jaws. It breathes underwater by trapping a jacket of air beneath its wingcases before diving. At night, the beetle may spread its wings and fly from one waterway to another, using moonlight to guide its way.

Longhorn beetles: These beetles are named after their antennae, which in some species are even longer than the body – not that this prevents them from flying. One longhorn, the titan beetle, is one of the largest of all insects, at more than 165 mm long. Pictured here is the rare little rosalia longhorn of Europe.

Tolype moth: Moths are hairy, right? But have you ever seen one quite this hairy? You may see the grey caterpillars of the North American tolype moth infest fruit trees in spring. The super-furry adult then flies in search of a mate during summer and autumn.

Cuckoo wasps: There are some 3,000 species of these beautiful wasps, which lead a solitary life, mainly in hot desert regions. Also known as jewel wasps – for obvious reasons – they get the name 'cuckoo' from the female's trick of laying her eggs in the nest of another species. The larva, when it hatches, eats any other larvae in the nest and is then fed by the unlucky host adult.

Flying insect record-breakers

One of the largest locust swarms on record covered a 200-km^2 area of Kenya, Africa, in 1954. It was estimated to contain 10 billion locusts.

What's the world's biggest moth? That depends on what you're measuring. The Atlas moth *Attacus atlas* and hercules moth *Coscinocera hercules* have the greatest wing area, but not the widest wingspan. That award probably goes to the white witch *Thysania agrippina* (above), an American species spanning up to 289 mm.

The painted lady *Vanessa cardui* is not only one of the world's most widespread butterflies, but its populations also make an epic migration each year. In spring, as their southern habitat heats up, they head north as far as the Arctic. Then in fall, as northern temperatures cool, they head south again, completing a round trip of up to 15,000 km. But no butterfly ever makes the full trip; instead, the migrant swarms go through several generations on each leg. In human terms, it would be like setting off for your annual holiday, but dying on the way, with your children's children's children finally arriving at the resort.

The world's toughest moth is possibly the Arctic woolly bear *Gynaephora groenlandica*, which lives in the polar north. Its life cycle from egg to adult takes around seven years… partly because the shaggy caterpillar spends more than 10 months of the year frozen, which leaves very little time for feeding! Uniquely among insects, the caterpillar can survive temperatures below −60 °C, but to do so it must pack its body with natural antifreeze chemicals before winter kicks in.

Glossary

abdomen the hind part of an insect's three-part body.

evolution developmental change, from one generation to the next, in all plants and animals. The change is driven by environment. For example, if certain flower throats become longer, the insects that feed on those flowers are likely to evolve a longer proboscis.

invertebrate animal that has no backbone, such as an insect, spider, worm or crustacean.

larva (plural: larvae) juvenile that hatches from the egg and later transforms into a pupa or, more directly, into an adult.

lek a gathering where animals choose a mate, often by showing off their strength or size. It's common among some birds – but paper wasps do it, too.

millisecond one-thousandth of a second.

parasite a life form that spends all or part of its life cycle on or inside another life form, which is known as the host.

pollinate the fertilisation process in plants, involving the transfer of male cells (in pollen) to female parts. This transfer can be accomplished by the wind, or by animal visitors such as birds, insects and bats. These animal agents are called pollinators.

predator an animal that kills and eats other animals. Roughly one-third of all insect species are predatory.

proboscis a tubular mouthpart in many insects, such as flies and butterflies, for taking in food, often by sucking.

pupa (plural: pupae) a stage of metamorphosis in which the insect larva rests inside a case and gradually transforms into an adult.

social forming a society in which individuals work together for the good of the group. Social insects include ants, termites, wasps and bees.

thorax the middle part of an insect's three-part body.

Index

The Author

British-born Matt Turner graduated from Loughborough College of Art in the 1980s, since which he has worked as a picture researcher, editor and writer. He has authored books on diverse topics including natural history, earth sciences and railways, as well as hundreds of articles for encyclopedias and partworks, covering everything from elephants to abstract art. He and his family currently live near Auckland, New Zealand, where he volunteers for the local Coastguard unit and dabbles in art and craft.

The Artist

Born in Medellín, Colombia, Santiago Calle is an illustrator and animator trained at Edinburgh College of Art in the UK. He began his career as a teacher, which led him to deepen his studies in sequential art. Santiago founded his art studio Liberum Donum in Bogotá in 2006, partnering with his brother Juan. Since then, they have dedicated themselves to producing concept art, illustration, comic strip art and animation.

Picture Credits (abbreviations: t = top; b = bottom; c = centre; l = left; r = right)
© www.shutterstock.com:

1 bl, 2 cl, 4 c, 6 tr, 6 bl, 7 tr, 7 bc, 9 c, 11 c, 13 c, 15 c, 17 c, 19 c, 23 c, 25 c, 27 c, 28 tl, 28 bc, 29 tl, 29 br, 32 cr.

21 © Richard Bartz, Munich (Wiki Commons)

索引

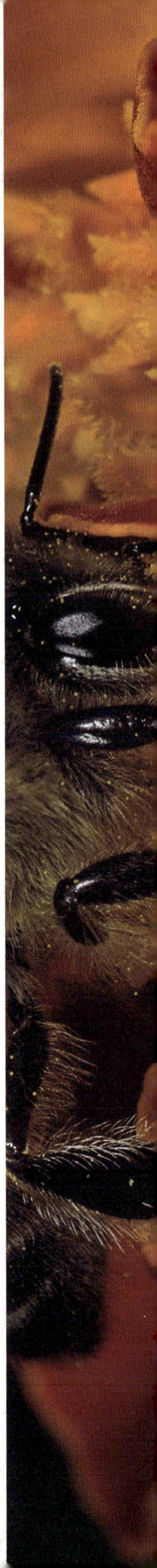

作者简介

马特·特纳出生于英国，20世纪80年代毕业于拉夫伯勒大学艺术学院，毕业后一直担任图片研究员、编辑和作者。他的书题材广泛，涉及博物学、地球科学和铁路等，并为百科全书和分册出版的丛书写过数百篇文章，从大象到抽象艺术无所不包。他现在和家人住在新西兰奥克兰附近，他还是当地海岸警卫队的志愿者，平时也涉猎工艺品的制作。

绘者简介

圣地亚哥·卡列出生于哥伦比亚的麦德林，是一位插画师和动画师，曾在英国爱丁堡艺术学院接受过培训。他的第一份职业是教师，教学促使他在连续性艺术领域继续深造。2006年，他和兄弟胡安在波哥大合伙创立了一家名为“自由德南”的艺术工作室，自此两人便致力于概念艺术、插画、连环画和动画的创作。

术语表

捕食者 捕食其他动物的动物。大约三分之一的昆虫都是捕食者。

传粉 植物的受精过程，具体指花粉中的雄细胞转移到雌性器官的过程。植物既可借助风媒传粉，也可由鸟类、昆虫、哺乳动物如蝙蝠等完成传粉。这些动物传播媒介被称为传粉者。

腹部 昆虫身体分为头、胸、腹三部分，腹部是最后面的一部分。

毫秒 千分之一秒。

喙 许多昆虫，如蝇、蝴蝶等，用来进食的管状口器（一般通过吸吮）。

寄生虫 在全部或部分生命周期内，寄生于另一个生物体内或体表的生物，被寄生的生物为其宿主。

进化 动植物在世代之间的发展变化。变化是受环境驱动的。例如，如果某种花的冠喉变长，那么吸食这种花蜜的昆虫就可能进化出更长的喙。

求偶比拼 动物聚集在一起，通过炫耀它们的力量或体形大小来求偶交配的活动。这种行为在鸟类中很常见，造纸胡蜂也有这种行为。

社会性 个体为了集体的利益共同工作，形成社会集群。社会性昆虫包括蚂蚁、白蚁、胡蜂和蜜蜂。

无脊椎动物 没有脊柱的动物，比如昆虫、蜘蛛、蠕虫、甲壳动物。

胸部 昆虫身体分为头、胸、腹三部分，胸部是中间的部分。

蛹 变态发育中的一个阶段。在这个阶段，昆虫的幼虫在蛹壳里逐渐成长为成虫。

幼虫 从卵孵化出来的幼体，在幼虫期结束后会化蛹或直接发育成成虫。

飞虫纪录创造者

1954年，非洲肯尼亚暴发蝗灾。这次蝗灾是有史以来规模最大的蝗灾之一，蝗群覆盖200平方千米，据估飞蝗的数量达到100亿只。

世界上最大的蛾子有多大？这取决于你的测量方式。

如果按翅膀的面积来算，最大的要数乌桕大蚕蛾和赫尔克里斯大蚕蛾。如果按翼展来算，冠军称号应该授予上图中的白女巫蛾，它是美洲的一种蛾，翅膀展开可达289毫米。

小红蛱蝶不仅是世界上分布最广的蝴蝶之一，而且每年都会进行漫长而艰难的种群迁徙。春天，随着南方的栖息地升温，它们向北迁徙，远至北极圈。到了秋天，北方的天气变冷，它们将再次向南迁徙，完成一次往返旅行，飞行距离长达15,000公里。

当然，没有任何一只蝴蝶完成过整个往返旅行，每一段迁徙之路都需要经过几代蝴蝶的努力。打个比方，这好比一个人动身去休年假，但是在半路去世了，最终孩子的孩子的孩子到达了度假地。

世界上最不屈不挠的蛾子恐怕要数生活在北极地区的北极灯蛾。在它的生命周期里，从卵到成虫需要经历大约七年。之所以要花这么长的时间，部分原因是毛茸茸的北极灯蛾幼虫一年中有十个多月的时间都处于冰冻状态，给进食留下的时间太少了。北极灯蛾毛虫可以在-60℃的低温下存活，这在昆虫世界里可谓独一无二。但为了生存，在入冬前，它就要把自己的身体用天然的抗冻物质包裹起来。

鸟巢枯叶蛾

蛾子毛多，对吗？不过，你见过长了这么多毛的蛾子吗？你或许在春天见到过大肆侵害果树的灰色毛虫，它们就是北美鸟巢枯叶蛾的幼虫。夏秋时节，超级多毛的成虫会飞出去寻找配偶。

杜鹃蜂

杜鹃蜂大约有3,000个品种，外观漂亮，独栖，主要生活在炎热的沙漠地区。杜鹃蜂也被称为宝石蜂。其之所以得名杜鹃蜂，很显然是因为雌蜂耍着和杜鹃鸟一样的伎俩，将卵产在其他蜂类的巢里。杜鹃蜂的幼虫孵化后，会吃掉宿主的幼虫，并由倒霉的宿主成虫喂食。

其他飞虫

世界上有太多漂亮的、令人着迷的昆虫，一本书远远不足以介绍完全。下面再列举一些有特点的飞虫。

龙虱

这种既庞大又勇猛的捕食者生活在淡水中，用尖利的颚抓捕其他昆虫、小鱼和蝌蚪。在潜水之前，龙虱会在鞘翅下面储存一层空气，供其在水下呼吸使用。晚上，龙虱会展开翅膀，在月光的指引下，从一片水域飞往另一片水域。

天牛

天牛的英文名称是“longhorn beetle”，因有一对长长的触角而得名。某些种类的天牛，其触角甚至比身体还要长，但是这并不妨碍飞行。泰坦大天牛是体形最大的昆虫之一，体长超过165毫米。下图所示是欧洲稀有的丽天牛。

剑角蝗

剑角蝗实际上并没有长“角”，只不过是长了一对短短的触角，这与它们的近亲蝈蝈以及灌木蟋蟀不同，后两者的触角比较长。剑角蝗大约有10,000多种，其中包括声名狼藉的飞蝗，它们能集结成数以百万计的蝗虫大军，在短短几个小时内把一片庄稼地毁灭殆尽。蝗虫拥有完美的飞行技术，不过相比之下，它们运动员级的跳跃技能更广为人知。

剑角蝗

寿命：成虫7—8周

大小：体长最高可达150毫米，平均体长为38—50毫米

“贪吃虫”

据估计，一吨重的飞蝗群在一天之内能吃掉2,500人吃掉的谷物。《圣经》里就描述了蝗虫吞噬埃及人粮食的灾难场面。

弹射器

蝗虫的后腿几乎就是一双弹射器，可以快速产生弹跳所需的巨大能量。

昆虫运动员

蝗虫在起跳时会收缩每条后腿的股节（位于足上部，较粗）处的两块肌肉，使其在相互牵拉中积聚大量的应变能（物体变形过程中贮存在物体内部的势能，译注）。接着，蝗虫会突然让股节上的其中一块肌肉放松，把储存的能量释放出来，然后胫节（位于足下部，较细）向下一蹬，将自己弹射到空中。

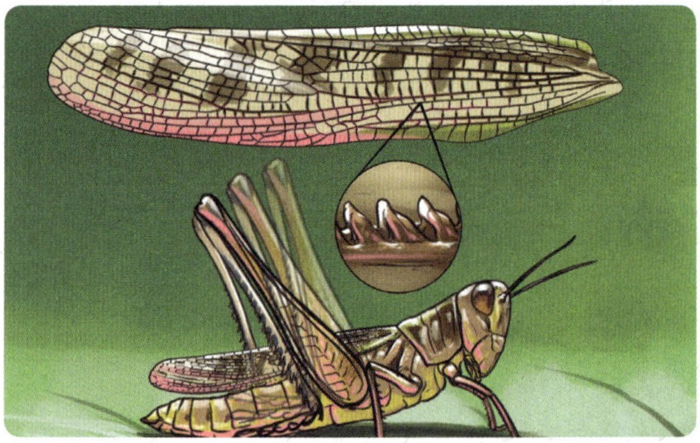

为了吸引配偶，蝗虫用后腿摩擦翅脉上的凹槽，发出唧唧的声音，这种现象被称为“摩擦发音”。

有些蝗虫色泽鲜艳，这是为了警示鸟类小心食用，免得坏了胃口。另一些蝗虫（如上图）会突然展开带有图案的翅膀，惊吓捕食者，赢得逃生的时间。

环境变化会引发蝗虫的幼虫和成虫成群地聚集、迁徙，毁坏庄稼，甚至会引发蝗灾地区的饥荒。

色彩鲜艳的蝗虫通常有毒，但也有一些蝗虫可以食用，并且富含蛋白质。据说，全世界大约有1,400种昆虫被列入了人类的食谱。

蚊子

这些外表柔弱的小飞虫属于寄生虫。雌蚊吸食宿主的血液，为腹中的卵提供营养。有些种类的蚊子还会传播疾病，每年导致至少200万人丧生，堪称世界上“头号夺命动物”。蚊子可谓精于“蚊道”，与其他昆虫相比，它们对声音更加敏感，为了寻找配偶会发出难听的嗡嗡声。

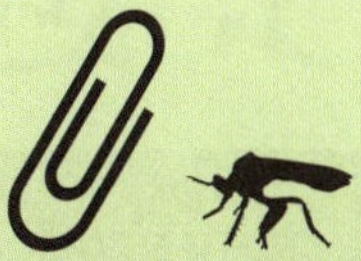

蚊子

寿命：少则几天，
多则1个月或更长时间
大小：体长2—19毫米，
平均体长为3—6毫米

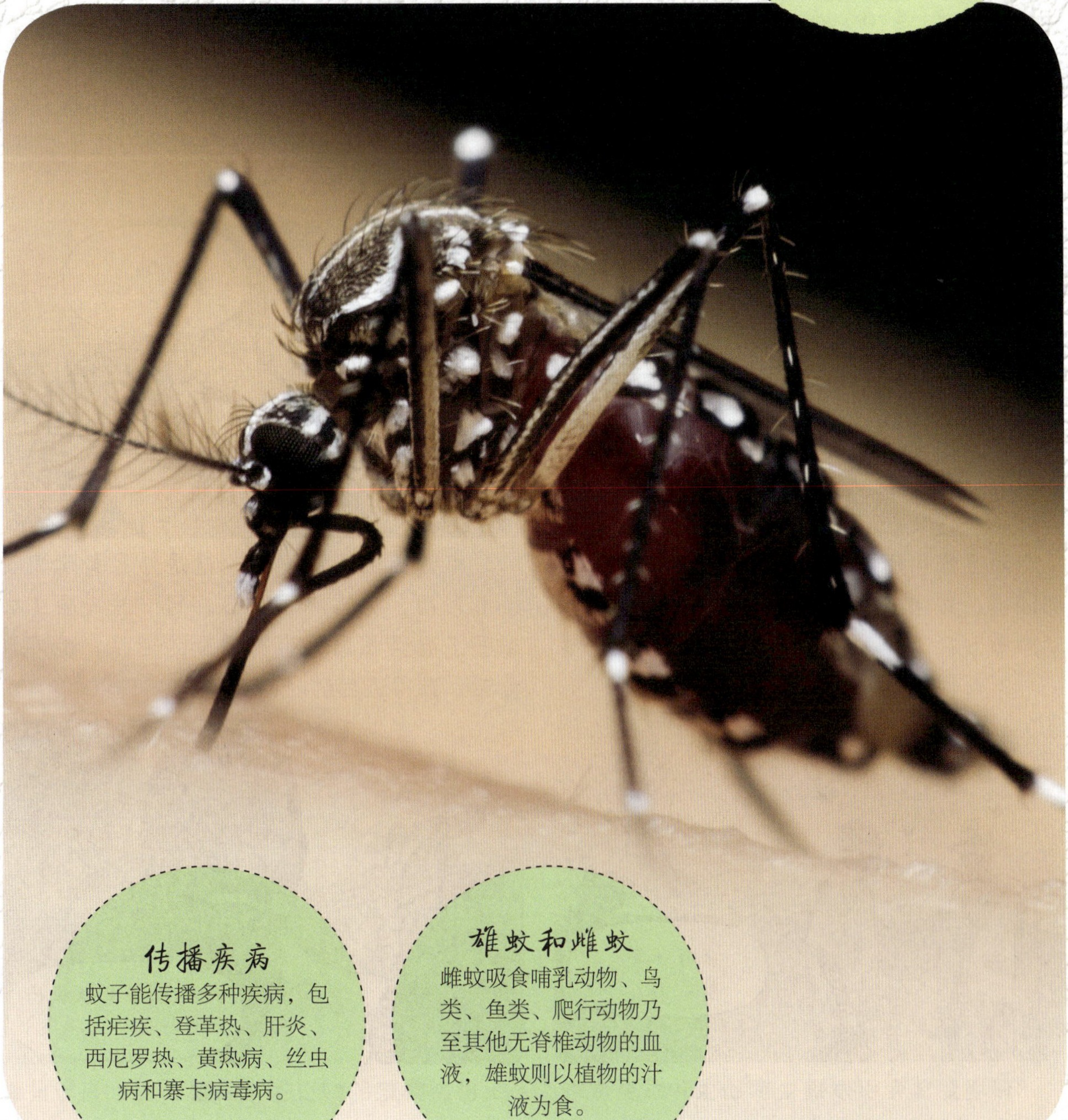

传播疾病

蚊子能传播多种疾病，包括疟疾、登革热、肝炎、西尼罗热、黄热病、丝虫病和寨卡病毒病。

雄蚊和雌蚊

雌蚊吸食哺乳动物、鸟类、鱼类、爬行动物乃至其他无脊椎动物的血液，雄蚊则以植物的汁液为食。

外表柔弱的夺命杀手

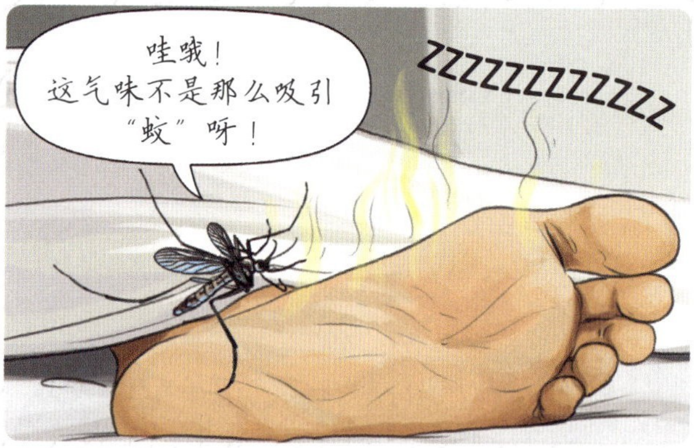

雌蚊既可以通过视觉，也可以通过嗅觉对宿主进行目标定位，哪怕是远在50米之外的汗馊味、呼出的气味等等，都能吸引它。

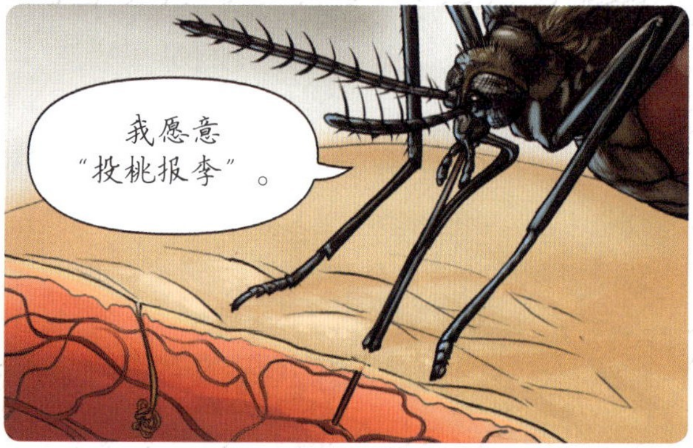

雌蚊吸血时将唾液注入宿主体内，唾液可以阻止宿主的血液凝结，确保口器顺利吸食血液。如果雌蚊的唾液中感染了病菌，就会传播疾病。

雄蚊触角上长有浓密的毛，可以帮助它听到未交配的雌蚊振动翅膀时发出的特殊嗡嗡声。如果雄蚊和雌蚊能够配对，那么两只蚊子就会“琴瑟和鸣”，就像是两个人哼唱同一段曲调，然后找到对方，仿佛是内置了“相亲”APP！

科学家花了很多年才发现蚊子会传播疾病。其中，医学先驱帕特里克·曼森（1844—1922）在他的园丁身上进行了疟蚊实验。

大多数蚊子的幼虫在水里发育，在靠近水面处呼吸，以水里的微小生物为食，并在靠近水面处化蛹后羽化成成虫，然后飞出去寻找配偶。

食虫虻

不是所有的蝇类都以糖、便便和腐肉为食。食虫虻科包括7,000多种食虫虻，遍布世界各地，个个都是飞得快、身子壮的昼行捕食者。它们足部强健，善于拦截飞行中的猎物，用如同针一样锋利的口器戳穿猎物的身体，注入唾液麻痹猎物，分解猎物的内脏。难怪它们被称为“刺客蝇”！

食虫虻

寿命：1年或2年

大小：体长3—80毫米

防治害虫

食虫虻捕食蚂蚁、甲壳虫、黄蜂、苍蝇、蝗虫、臭虫，等等——实际上，凡是会动弹的虫子，没有它不吃的。所以说，食虫虻是防治害虫的得力助手。

吃鸟的食虫虻

人们发现，个头大的食虫虻居然会攻击蜂鸟，这其中有被称为“蜜蜂猎豹”的红足食虫虻和被称为“食蜂恶魔”的绒毛食虫虻。

俯冲的猎人

食虫虻常在阳光灿烂的日子里捕食。它们栖息在植物上伺机而动，凭借完美的视力和顶级的飞行技能，瞄准猎物的飞行路线，在飞行中快速攻击目标。它们的腿不但健壮，而且带刺，这有助于它们抓握猎物——它们总是选择足够小的猎物，以便能够轻易地抓住。

许多食虫虻进化出了具有警戒性的色彩和外观，帮助它们躲避捕食者。比如，上图中左边这只食虫虻和右边的这只蜜蜂看上去像极了。

食虫虻的口器上方长有胡子一样的刚毛，称为"口鬃"。在猎物挣扎时，口鬃可以保护眼睛不受伤害。

食虫虻将有毒的唾液注入猎物体内，用以麻痹猎物，将猎物的内脏软化后吸食。一只大个头的食虫虻可以用它强有力的口器穿透甲壳虫的翅鞘。

有一种食虫虻的雄性个体后腿上长有特殊的羽状物，可能用来在求偶"舞蹈"中吸引雌性的眼球。

绿头苍蝇

蝇是如何飞行的？在双翅目蝇科昆虫中，飞行的秘密在一定程度上就藏在平衡棒上。你可以阅读左页的内容来了解这个巧妙的身体构造是如何工作的。同时，具有金属光泽的绿头苍蝇是一种害虫，它们一会儿趴在马路上刚刚被车辆撞死的动物尸体上，一会儿又蹿进厨房，趴到多汁的牛排上产卵并传播疾病。它们的幼虫就是经典恐怖电影里的蛆。其实，蛆在医学上是有用武之地的，保准让你大吃一惊。

绿头苍蝇

寿命：成蝇通常最多存活1个月，如果能安然度过冬眠，则可以活得更久

大小：体长10—14毫米

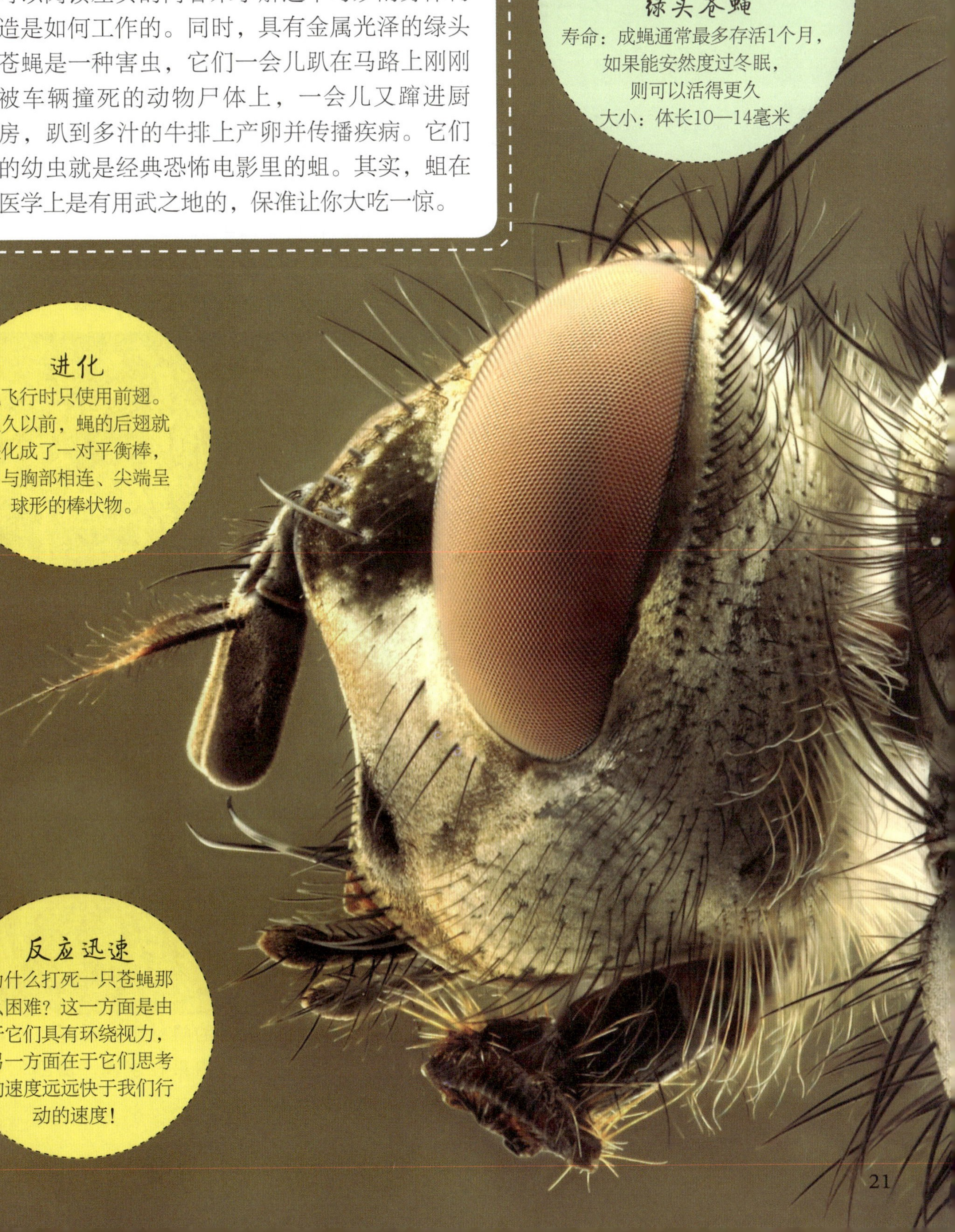

进化

蝇飞行时只使用前翅。很久以前，蝇的后翅就进化成了一对平衡棒，即与胸部相连、尖端呈球形的棒状物。

反应迅速

为什么打死一只苍蝇那么困难？这一方面是由于它们具有环绕视力，另一方面在于它们思考的速度远远快于我们行动的速度！

快如闪电

在飞行中，苍蝇通过振动平衡棒来检测身体在俯仰、翻滚和偏航动作（上下左右移动）中的变化。平衡棒会向胸部的神经组织发送信号，不断调节飞行肌，保持身体的平衡和稳定。如果你想观察平衡棒，可以找只大蚊看一下（大蚊模样像蚊子，但并不咬人，是蚊子的远房亲戚，译注），大蚊的那对平衡棒可大了。

当你试图拍一只苍蝇时，苍蝇会计算危险靠近的角度，并在0.01秒内调整腿的方向，让自己沿着安全的方向跳离。

绿头苍蝇往往是第一拨在尸体上产卵的苍蝇，这是因为生出的蛆吃腐肉。刑侦专家通过寻找尸体上的绿头苍蝇来计算死者的死亡时间。

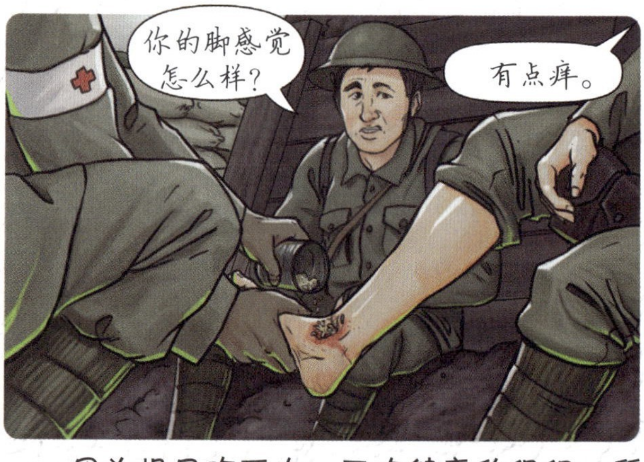

因为蛆只吃死肉，不吃健康的组织，所以人们用蛆来清理溃烂的伤口，比如在战争中用蛆清理受伤士兵的伤口。

科学家正在研究绿头苍蝇的卵——蛆体内的各种天然化学物质，这可能有助于研发出治疗耐药菌感染，甚至癌症的新药。

蜻蜓

蜻蜓是出类拔萃的飞行能手，能够向下俯冲，追捕空中的猎物。这一方面得益于它们的环绕视力，另一方面得益于其胸部强大的飞行肌与翅膀根部直接相连的生理构造。欧洲最大的蜻蜓是帝王伟蜓，它能够在湖泊和河流上空连续捕猎好几个小时，尤其是在阳光充足、能让飞行肌保暖的天气里。

帝王伟蜓

寿命：成虫通常可以存活4周，最多可达8.5周

大小：体长66—84毫米，翼展平均为106毫米

宽大的翼展

蜻蜓和豆娘一起组成了蜻蜓目。它们当中体形最大的是中美洲的巨豆娘（也叫“直升机豆娘”），翼展可达191毫米。

想飞哪里飞哪里

蜻蜓在追赶猎物或竞争对手时，能朝任意方向飞行，甚至能倒立着飞行。但是由于前腿的结构，蜻蜓不能行走。

飞行大咖

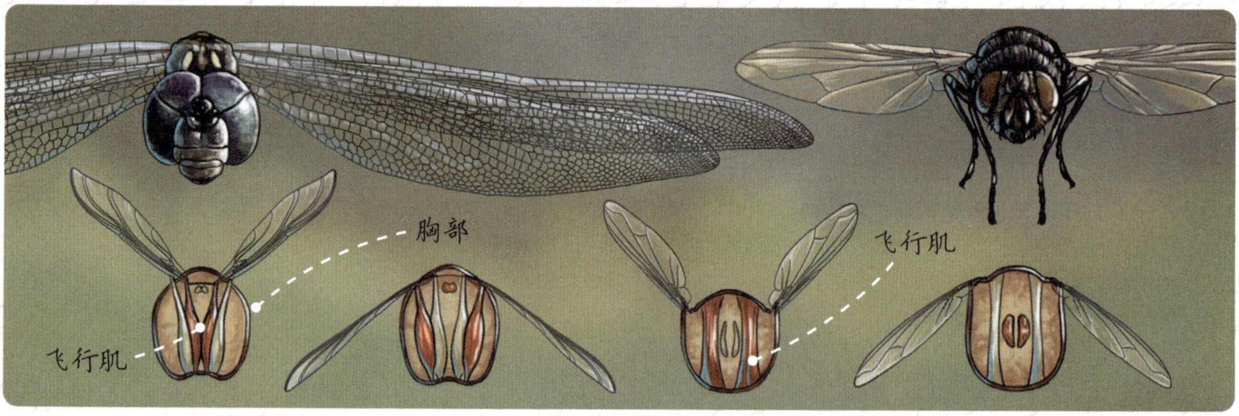

蜻蜓的翅膀虽然排列方式很原始（上图左），但强健有力。蜻蜓的四只翅膀根部各与一块飞行肌相连，以便实现对翅膀直接的、一对一的控制。而像家蝇（上图右）这样更加现代化的昆虫则会采取非直接控制的飞行方式，即通过让胸部变形来收紧飞行肌，正如上图中的横截面所示。

蜻蜓长有一对巨大的复眼，两只眼睛一般在头顶上连到一起，从而使蜻蜓具备了“环绕视力”。为了获得更好的双眼视觉，一些蜻蜓的眼距很宽。

蜻蜓的稚虫会在池塘底部度过两年的光阴。它属于捕食性动物，能在不到25毫秒的时间内快速伸出脸盖（收合起来的下唇）将猎物擒获。

雄蜻蜓具有领地意识，会在空战中驱逐对手，保卫池塘中属于它的那一小片天地。澳大利亚帝王伟蜓具有“运动伪装”的绝技，能选择一条特殊的飞行路线，让对手误以为自己是静止不动的。这样一来，对手就没法儿将它从背景景物中辨认出来了。

造纸胡蜂

大约在1980年，人们在美国发现了欧洲造纸胡蜂。从那以后，欧洲造纸胡蜂在全美建立了根据地，它们将当地的造纸胡蜂赶走，猎捕昆虫喂养后代。造纸胡蜂的巢穴可谓是工程学上的奇迹：蜂房开放朝外，聚成球形，常由植物的细茎吊起整个蜂巢。造纸胡蜂经常成群结队地在花园里出没。尽管不及黄胡蜂那么凶猛，但是如果你离得太近，它们也会蜇你。

欧洲造纸胡蜂

寿命：蜂王可以存活数年，工蜂和雄蜂11个月

大小：体长8.5—13毫米

“纸质”蜂巢

造纸胡蜂是以它们“纸”质的筑巢材料来命名的。这种材料实际上是造纸胡蜂嚼碎的木材和唾液混合形成的糊状物。

能防水的唾液

造纸胡蜂的唾液具有意想不到的防水功能，科学家仿照其唾液的成分，为无人驾驶飞行器开发出了一种特殊的防水涂层材料。

专业工程师

造纸胡蜂在飞行时，长长的后腿会垂下来拖着，我们可以通过这个特征来鉴别它。相比之下，一只普通黄胡蜂的后腿要“利落”得多。

造纸胡蜂常从旧栅栏上收集木材来筑巢。仔细听，你能听到微弱的嘎吱声，这就是它们啃咬木头的声音。

在繁殖期，雄蜂和雌蜂会聚集起来进行“求偶比拼”。在这场比拼中，雄蜂会竞相争抢高处的栖息区，你拥我挤，你追我打。当然，颜值也很重要：面部黑斑最多的雌蜂和腹部斑纹最齐整的雄蜂被认为是最理想的伴侣。

造纸胡蜂吃的昆虫五花八门。夏天，你甚至能看到它们捕食毛毛虫，因为毛毛虫可以为成长中的幼虫提供蛋白质。

寄生性“杜鹃造纸胡蜂”自己不会筑巢，而是占用其他蜂的蜂巢，逼迫巢里的工蜂喂养她的幼虫。这一行为和本书第14页介绍过的“杜鹃大黄蜂”类似。

大黄蜂

大黄蜂胖乎乎、毛茸茸的，能为野生植物和农作物传粉，是受人欢迎的花园访客。大黄蜂同蜜蜂一样，是社会性昆虫，但是蜂群规模要小得多，通常一个蜂群中至多有大约400只大黄蜂。与蜜蜂不同的是，大黄蜂能够在寒冷的天气中飞行，所以它们大多生活在气温较低的地区。但这也意味着到了秋天，蜂群里的大黄蜂通常会相继死去。等到春天来临，年轻的蜂王需要重新建立蜂巢。

黄尾大黄蜂

寿命：工蜂4—7周，蜂王1年

大小：工蜂体长11—17毫米，蜂王20—22毫米

大黄蜂的体重

空气动力学"专家"曾经推断说，大黄蜂身体太重，没法儿飞起来！但是，这些专家忽略了一个事实，那就是在小小的大黄蜂与大飞机周围，气流是不一样的。

邓布利多

大黄蜂在古英语中也叫做"dumbledore"。"哈利·波特"系列的作者J.K.罗琳给霍格华兹魔法学校的校长取名叫"邓布利多"，就是用了这个词。或许，在罗琳的想象中，这位校长嘴里总在哼着小曲儿。

毛茸茸，叫嗡嗡

大黄蜂是苹果等春天开花的作物的重要传粉者，它的后足有毛茸茸的花粉篮，用来收集花粉。

越冬之后，每年春天，蜂王会独自建造一个新的蜂巢。它从腹部蜡腺分泌出蜂蜡，从第一个蜂房开始建起。

蜜蜂的蜂巢结构整齐有序，相比之下，大黄蜂的巢要么建在一堆乱七八糟的树叶上，要么建在树洞中，蜂房杂乱无章。这些蜂房用来存放蜂蜜、花粉，同时还用做育幼室。工蜂负责保洁，把死掉的大黄蜂挪走。

秋天，雌蜂（年轻的蜂王）与雄蜂交配。交配后，雄蜂死去，雌蜂大量进食，长得肥肥胖胖的，准备冬眠。

所谓的“杜鹃大黄蜂”是一种雌蜂，它入侵蜂巢，将原有的蜂王赶走（如上图），然后产下自己的卵，让巢中原有的工蜂照顾它们。

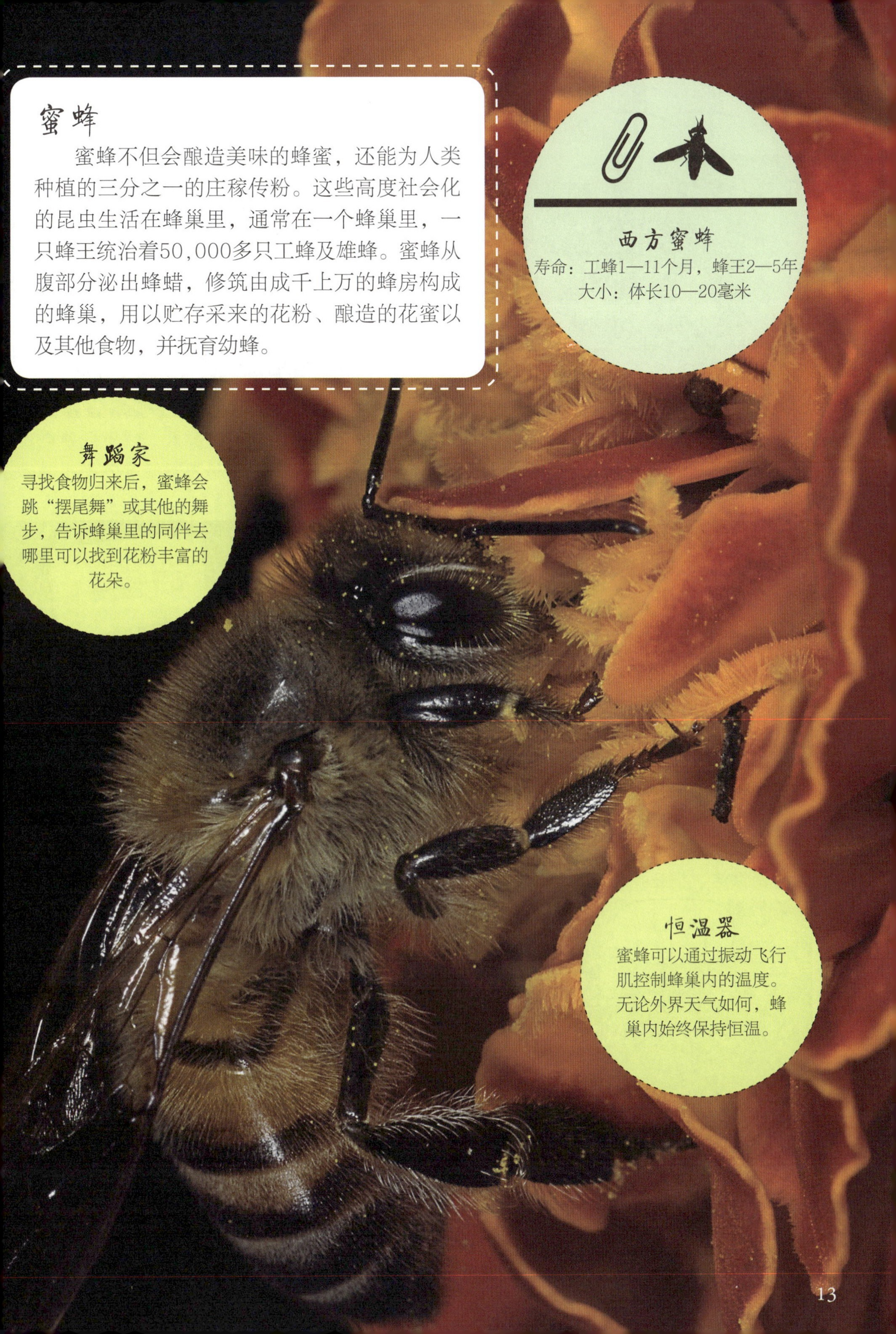

蜜蜂

蜜蜂不但会酿造美味的蜂蜜，还能为人类种植的三分之一的庄稼传粉。这些高度社会化的昆虫生活在蜂巢里，通常在一个蜂巢里，一只蜂王统治着50,000多只工蜂及雄蜂。蜜蜂从腹部分泌出蜂蜡，修筑由成千上万的蜂房构成的蜂巢，用以贮存采来的花粉、酿造的花蜜以及其他食物，并抚育幼蜂。

西方蜜蜂

寿命：工蜂1—11个月，蜂王2—5年

大小：体长10—20毫米

舞蹈家

寻找食物归来后，蜜蜂会跳“摆尾舞”或其他的舞步，告诉蜂巢里的同伴去哪里可以找到花粉丰富的花朵。

恒温器

蜜蜂可以通过振动飞行肌控制蜂巢内的温度。无论外界天气如何，蜂巢内始终保持恒温。

生活甜蜜蜜

我是最棒的！

蜂王（上图中）是蜂巢的建立者，一天可产多达1,500枚卵，受精卵孵化成工蜂（上图右）。工蜂负责寻找食物、清理蜂巢、抚育幼蜂（上图最右），而且还会蜇人！未受精的卵发育成雄蜂（上图左），专门负责与未来建立新蜂巢的年轻蜂王交配。

人工养殖的蜜蜂生活在人造蜂巢里，而它们的野生祖先则是在树洞或类似的岩缝里筑巢。

我喜欢
猜字谜！

通过跳“8字形”摆尾舞，一只蜜蜂以太阳的位置为基准，向蜂巢里的同伴指出蜜源、花粉源或水源的方向。

虽然它们脾气有点坏，
但离了它们还真不行。

被蜜蜂的巨型亚洲近亲排蜂蜇了会很疼，可农民们居然欢迎它们的到来。谁让它们能为棉花、芒果、椰子、咖啡树、辣椒等农作物授粉呢！

我是个“脏兮兮”的
小怪兽，蹦跶不了
几天啦。

蜜蜂是不可或缺的传粉者，但由于人们过量使用杀虫剂，以及近来寄生在蜂巢中的蜂螨迅速扩散，蜜蜂的生存面临着威胁。

蜂鸟鹰蛾

如果你发现一只巨大的蛾子在花丛上空一边盘旋，一边用细长如线的喙吸食花蜜，那么它很可能就是天蛾。天蛾家族中有北美白浆果透翅蛾（又称“飞行的龙虾”），还有欧洲和亚洲的蜂鸟鹰蛾。它们都进化出了超强的飞行技能。仔细听，你或许能听到它们拍打翅膀时发出的轻轻的嗡嗡声。

蜂鸟鹰蛾

寿命：7个月（含冬眠期）

大小：翼展40—45毫米

冬眠

在其分布的某些区域，蜂鸟鹰蛾会在树上或岩石上寻找舒适的隐蔽处冬眠，从十月一直睡到次年四月。

繁殖

雌性蜂鸟鹰蛾一年产卵多达四次，卵产在蓬子菜上，方便幼虫孵出后食用。

蜂鸣与悬停

蜂鸟吸食花蜜时可以快速扇动翅膀在空中悬停，频率高达每秒70—80次。蜂鸟鹰蛾也会这一招，故得此美名。

嗯——嗯——嗯
哦，不，朋友，我们那大名鼎鼎的嗡鸣可不是这样的。

天蛾指体形巨大、飞行能力强的蛾子。上图右侧的北美白条天蛾就是其中的一员，它也可以像蜂鸟一样在空中悬停。

我敢打赌，你可没这本事！

1862年，查尔斯·达尔文对马达加斯加的一种巨型兰花进行了研究，他很想知道是哪种昆虫在为这种植物传粉。1903年，人们找到了这种传粉的蛾子。瞧，它的喙可真长！

呼，
又没捉到！

蜂鸟鹰蛾是引人瞩目的飞行家，能够在悬停的同时“横向滑动”以躲避捕食者，比如上图中的这只螳螂。

有些蝇类有超长的喙。在不相关的动物类群中分别进化出相似的特征，这种现象叫做趋同进化。

鸟翼凤蝶

这种美丽的蝴蝶是一种凤蝶。它们不仅翅膀长，而且善于飞行，所以常常被比拟为鸟儿——这正是“鸟翼凤蝶”这个名字的来历。它们也是所有蝴蝶中体形最大的一种。在巴布亚新几内亚发现的亚历山大女皇鸟翼凤蝶拥有最宽的翼展，约为254毫米。鸟翼凤蝶很少受到捕食者的困扰，这要归功于它们饮食中所含的毒素。

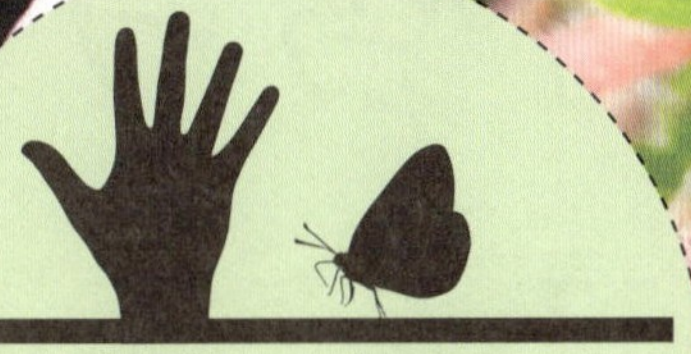

石冢鸟翼凤蝶

寿命：多达3个月

大小：体长70毫米，翼展125—150毫米

艳丽的雄蝶

红颈鸟翼凤蝶是马来西亚的国蝶，雄蝶的翅膀黑绿相间，比雌蝶更为艳丽显眼。其他种类的鸟翼凤蝶也是如此。

采集鸟翼凤蝶

由于鸟翼凤蝶在野外极为罕见，因此对它的采集有严格的法律规定。不过，大多种类的鸟翼凤蝶是可以笼养的。

超大的翅膀

哥们儿，放轻松，我只是来找点花蜜！

当一只雄蝶向一只雌蝶求爱时，雄蝶会在雌蝶上方盘旋，并向它释放类似香水气味的信息素。这种用化学物质求偶的方式可以促使雌蝶和雄蝶交配。雄蝶还会驱赶情敌，甚至把鸟撵走！

唷，你好臭啊！
你闻起来也不怎么清新啊。

鸟翼凤蝶的幼虫味道可不好，因为它们以有毒的藤本植物为食，毒素逐渐累积在身体组织内，直至成年。它们头部后侧还有一种能散发恶臭的器官，叫做丫腺，可以令负鼠等入侵者“敬而远之”（如上图）。

忽略我就好。

鸟翼凤蝶的蛹通过一根精细的丝线固定在小树枝上，看起来就像一片卷曲的树叶。这种绝妙的伪装帮助它隐藏起来，不被捕食者发现。

我觉得你味道还不错！

鸟翼凤蝶唯一的天敌是络新妇属金丝蛛，这种大型蜘蛛并不介意鸟翼凤蝶因体内毒素导致的糟糕味道。

本书介绍有翅昆虫，尤其是那些个头大、颜值高、飞行速度快、飞行技艺高或非常危险的昆虫。有些昆虫虽然不具备这些特性，但其种类繁多，令人叫绝，所以也会一并介绍。当然，有翅昆虫的种类远远不止书中提到的这些，你在自己居住的地方就能发现很多。

虫儿飞

最早的昆虫为什么飞上了天？你可能还会问，在大约1.5亿年前，植物开始开花，这又是为什么？这两个问题的答案都是：为了生存。事实上，植物正是利用芬芳艳丽的花朵，吸引昆虫和其他动物前来授粉。大约4亿年前，地球上出现了最早的有翅昆虫，那时候植物生长得越发高大，形成了最初的森林。因为有了翅膀，早期的昆虫才得以逃避捕食者，后来又得以像今天的蜜蜂、蝴蝶等昆虫那样，采食花粉和花蜜。

翅膀是昆虫的立身之本，因此，昆虫的主要种类都是根据翅膀的希腊语pteron来命名的。例如，蜻蜓和豆娘作为最早飞上天的昆虫，属于Palaeoptera（古翅下纲），意思是“古老的翅膀”。这些描述性的名字可以帮助我们辨别昆虫。又如，Coleoptera（鞘翅目）意为“变硬的翅膀”，这一目的昆虫又被称为甲壳虫，它们会把后面的一对翅膀合拢起来，收在具有保护作用的坚硬前翅下。蜉蝣属于Ephemeroptera（蜉蝣目），意思是“短命的、有翅的”。蜉蝣成虫的寿命短暂，通常仅仅能维持几个小时。

目录

飞虫拥有一大生存优势——翅膀。这一结构功能强大，将普普通通的虫子变成了出类拔萃的佼佼者，造就了惊艳的空中杂技。到这本书里，近距离了解：

- 蜂鸟鹰蛾如何一边进食，一边躲避捕食者
- 蜜蜂如何通过舞蹈指示蜜源的方向
- 帝王伟蜓如何在飞行中伪装自己

……

当然还有更多王牌飞行员的趣事。在这里，你还会了解到建筑大师胡蜂和蜜蜂、食肉飞虫食虫虻、伪装成叶子的蝴蝶蛹，还有惹人讨厌的蚊子……真让人眼花缭乱！读完这本书，相信你也会期待长有翅膀哦！

番茄博士双语百科

不可思议的虫子王国

身怀绝技的飞虫

马特·特纳（英）著
圣地亚哥·卡列（英）绘
丛岚 译

外语教学与研究出版社
FOREIGN LANGUAGE TEACHING AND RESEARCH PRESS
北京 BEIJING

Record-Breaking Bugs

SUPER SPIDERS

Written by Matt Turner
Illustrated by Santiago Calle

外语教学与研究出版社
FOREIGN LANGUAGE TEACHING AND RESEARCH PRESS
北京 BEIJING

With their unique ability to spin silk that is stronger than steel, spiders make cunning traps and trip wires as well as beautiful orb webs. This book brings you up close to these brilliant engineers, aerial acrobats and super successful predators. Discover the

- trapdoor spider that can pull 38 times its own weight
- diving bell spider that lives its whole life in water – but it needs air to breathe!
- spiders that look like bird droppings or beetles

...

and many more. This book will let you into some fascinating spidery secrets. So stop by and look next time you see a web glittering in the sun, and its busy architect dangling by a thread...

Contents

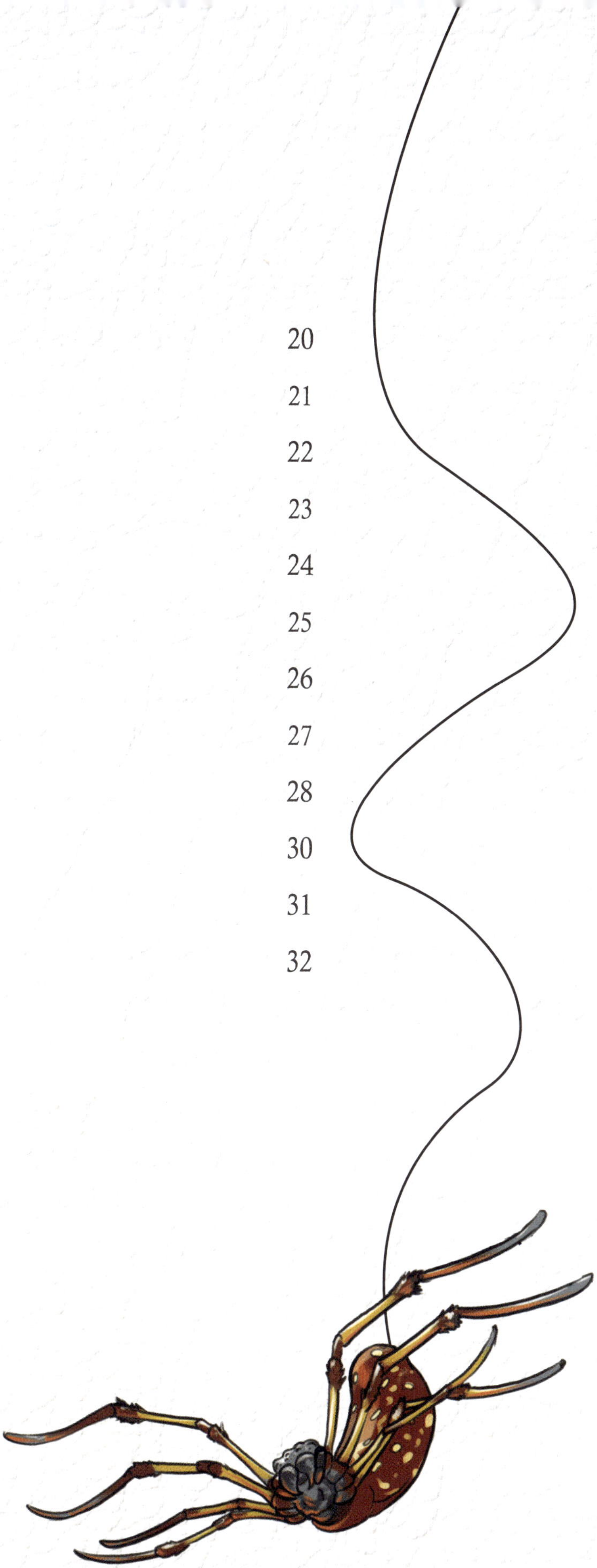

Spectacular Spiders

For many of us, spiders are fearsome creatures. But they're also fascinating. They've lived on Earth for more than 300 million years, and have evolved to live in almost every dry-land habitat, from deep caves to deserts and rainforests. Many seem quite at home in our houses. Today there are more than 50,000 species of all shapes and sizes, found worldwide except Antarctica.

Spiders are not related to insects; they have eight (not six) legs, their body is in two (not three) parts, and they lack antennae. Like insects, however, spiders are arthropods. They have an exoskeleton (outer armour), which must be moulted regularly as the young spider grows to adulthood. (You can sometimes find the empty moults in their webs.) All spiders make silk, a strong thread that is spun from the spinnerets – tiny nozzles at the tip of the abdomen. Most spider species have eight eyes, but some have six, or four, or even just two.

A spider's body has two main parts: the cephalothorax (supporting the eyes, mouthparts and legs); and the abdomen (containing most of the internal organs).

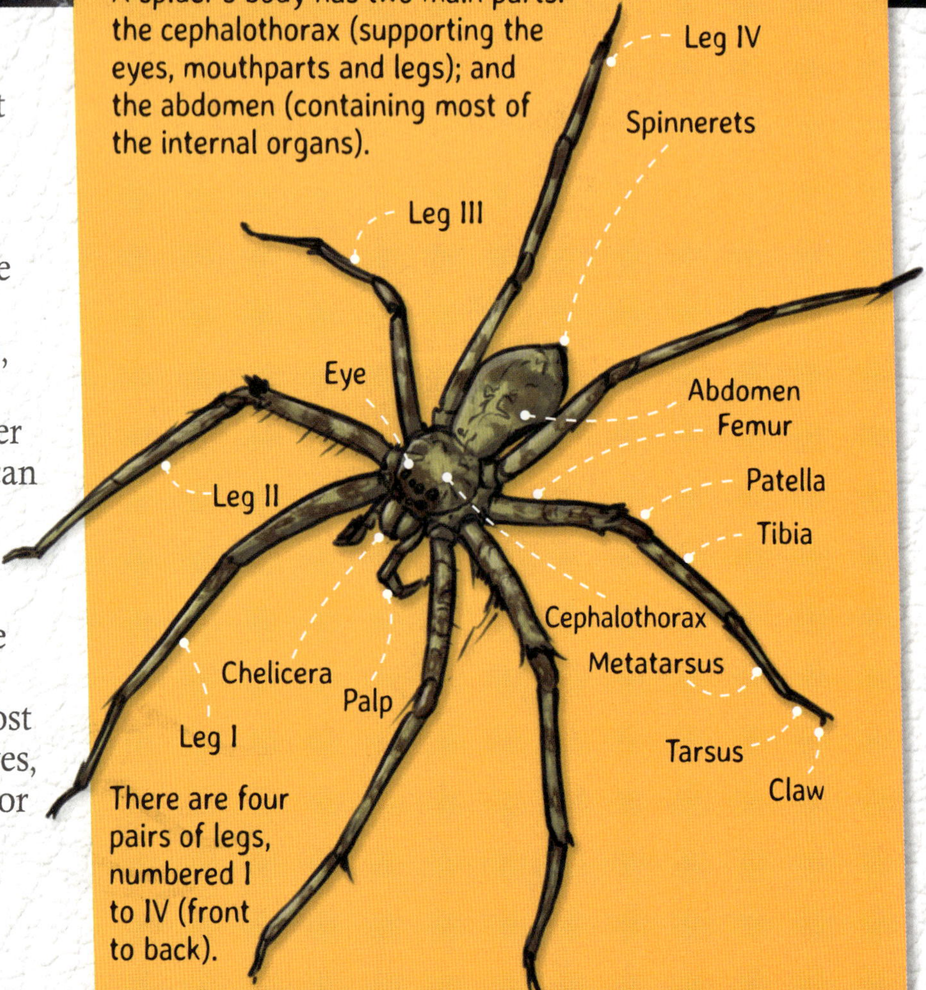

There are four pairs of legs, numbered I to IV (front to back).

Spiders are predators – some catch prey in a web, others ambush it, while still others chase after it. Most use venom (poison) to kill or paralyse their prey and turn its insides into a kind of 'gut soup', because spiders can't eat solid food. In some species, the venom is so potent it can harm or even kill humans; but the huge majority of spiders cannot hurt us – and even help us by killing pests. It's been estimated that the prey eaten each year by all the spiders weighs more than the entire human race.

Silk & Orb Webs

Some spiders have a cribellum – a rack of extra spinnerules – which spins a fine silk that is then combed by the legs to make 'woolly' silk for snaring prey.

Silk is stretchy and strong, and spiders often spin a 'dragline' when they leap or drop from a perch – such as a twig – to catch their fall.

Silk is so strong that, in earlier times, tribes in New Guinea wove it over wooden hoop-nets to catch fish from rivers.

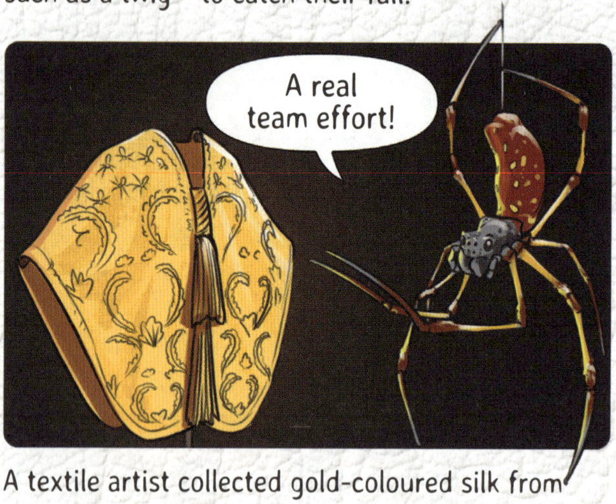

A textile artist collected gold-coloured silk from more than a million orb weavers to make this beautiful cape.

Stabilimenta are silken zigzags that some spiders weave into their webs. Do these eye-catching designs help attract insect prey by reflecting ultraviolet light, or do they keep birds away by making the web more visible? Or do they help to make the spider look bigger? Experts aren't quite sure.

MAKING SILK

Stored as a liquid, spider silk dries into a thread after being squeezed out of the spinnerets. Spiders can spin different kinds of silk: sticky for trapping prey, non-sticky for walking on, extra-strong for hanging from, and so on. Their most famous creation is the beautiful, spiralling orb web, but many spiders build more messy-looking, three-dimensional 'tangle webs'.

GOLDEN ORB WEB SPIDER

Body length: female up to 50 mm, male 5–8 mm
Where found: Australia, America, Africa and Asia

STRONG STUFF

Spider silk is five times stronger than steel of the same thickness. One day, it may be used to make bullet-proof vests for soldiers.

GARDEN SPIDERS

A garden spider may use up to 60 m of silk in a typical orb web, but usually finishes the task within an hour.

Tunnels & Trapdoors

Check out
my new digs.

A burrowing spider uses spines on its fangs to dig a hole in the soil. It then waterproofs the burrow walls with layers of mud and spit and then silk.

delivered to

A trapdoor spider adds a hinged door made of silk, mud and grasses. This hides the spider as it sits in wait for prey to wander near.

If I can't see you,
you can't see me.

With its door shut, the burrow is hidden from view. But spider-hunting wasps can usually spot the entrance. The door is no proof against floods, either.

Just my size...
'funnelly' enough.

Spiders in the family Agelenidae, which build sheet webs above ground, add funnel-like hideaways to their webs, from which they rush out to grab prey.

I'm not just
a pretty face,
you know.

The trapdoor spider *Cyclocosmia* has a flat-ended abdomen covered with a tough plate. It uses this to 'stopper' itself inside its burrow, if threatened.

WEB TRAPS

Ancient spiders lived in holes in the ground. Some spiders still live like this today, particularly the primitive species known as mygalomorphs. Trapdoor spiders often add hinged lids to hide their entrance. Funnel-web spiders, by contrast, ring their entrances with silken 'trip wires' that alert them when prey is walking nearby. Above the ground, some spiders add protective tunnels of silk to their webs.

BANDED TUNNEL WEB SPIDER
Body length: up to 20 mm
Where found: New Zealand

SUPER STRONG
Using its barbs and fangs to hold its trapdoor closed, the California trapdoor spider *Bothriocyrtum californicum* can resist a record-breaking pull of up to 38 times its own weight.

MASTER DIGGERS
Trapdoor spiders can dig tunnels 30–40 cm deep in the ground, and live up to 30 years.

VENOM & HUNTING

Spiders' fangs work in one of two ways. The fangs on mygalomorphs (suborder Orthognatha) work up and down together, a bit like a pair of pickaxes...

... whereas in the so-called 'true' spiders (suborder Labidognatha), the fangs open and close in a side-to-side pincer action.

Trapdoor spiders are sit-and-wait predators. When prey walks near, they burst out and grab it, then pull it back into their lair.

When an insect blunders into an orb web, it takes only 5–10 seconds for the spider to rush out and bite it. Then it wraps the prey in silk before eating it.

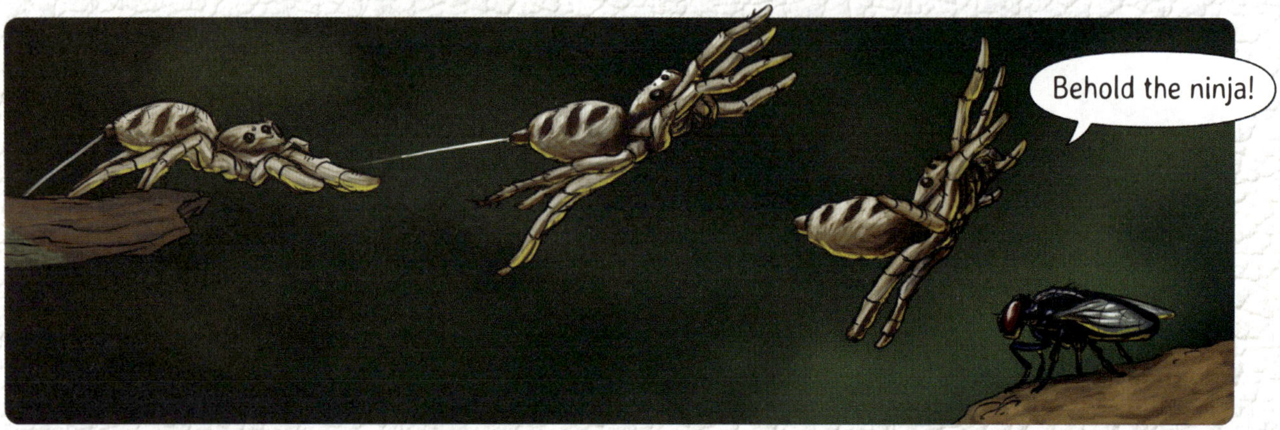

Jumping spiders pounce on their prey like tigers. There are more than 5,000 species of jumping spider worldwide. Two of their eight eyes, the median pair, are especially large and look directly forward, helping them judge accurate distances over a range of several centimetres. (It also makes them look rather cute!)

TOXIC BITES

Almost all spiders rely on venom for hunting: one quick bite and their dinner stops struggling! When a spider stabs its fangs into prey, powerful venom flows out through the hollow fangs, causing paralysis or death, so the spider can feed at leisure. The venom softens the victim's insides into a liquid 'soup', which the spider then sucks out. But first, of course, a spider has to catch its prey.

WOLF SPIDER

Body length: 10–35 mm
Where found: worldwide except Antarctica

FANGS

The woodlouse spider specialises in eating – can you guess? – woodlice. It has very strong fangs for piercing their exoskeleton.

VENOM

There are two main venom types. Neurotoxins attack the prey's nervous system and stop it moving. Cytotoxins dissolve the guts. Spiders may have one type or the other, or a mixture.

Throwing & Spitting

The net-caster first spins frame-lines as a support structure. It then spins a small, net-like web onto its hind legs. The net silk is very stretchy.

The spider now hangs downwards from its hind legs and holds the little 'net' in its front four legs, ready to pounce.

When a beetle walks into range, the spider stretches the net out by up to 10 times its original size, and casts it over the prey. The silk, spun using the spider's cribellum (see page 8), is made up of strands so crinkly and fine that the prey becomes completely entangled in it.

The spitting spider squirts twin jets of silky venom from its fangs. Meanwhile, its body vibrates from side to side to create zigzag patterns in the silk, which glues the prey down like a sticky net. The attack is so fast (about three-hundredths of a second) that it can only be seen with a slow-motion camera.

AMBUSH PREDATORS

Net-casting spiders live in warm habitats from South America to Malaysia and Australia. They spin a little web that is just like a net, wait in ambush, then throw it down over a passing victim. They use their superb night vision – 12 times better than ours – to locate prey in darkness. Spitting spiders, which are more or less worldwide, squirt a mix of venom, glue and silk at their prey to pin it down.

NET-CASTING SPIDER

Body length: female up to 18 mm, male up to 14 mm
Where found: Australia, Asia, Africa and America

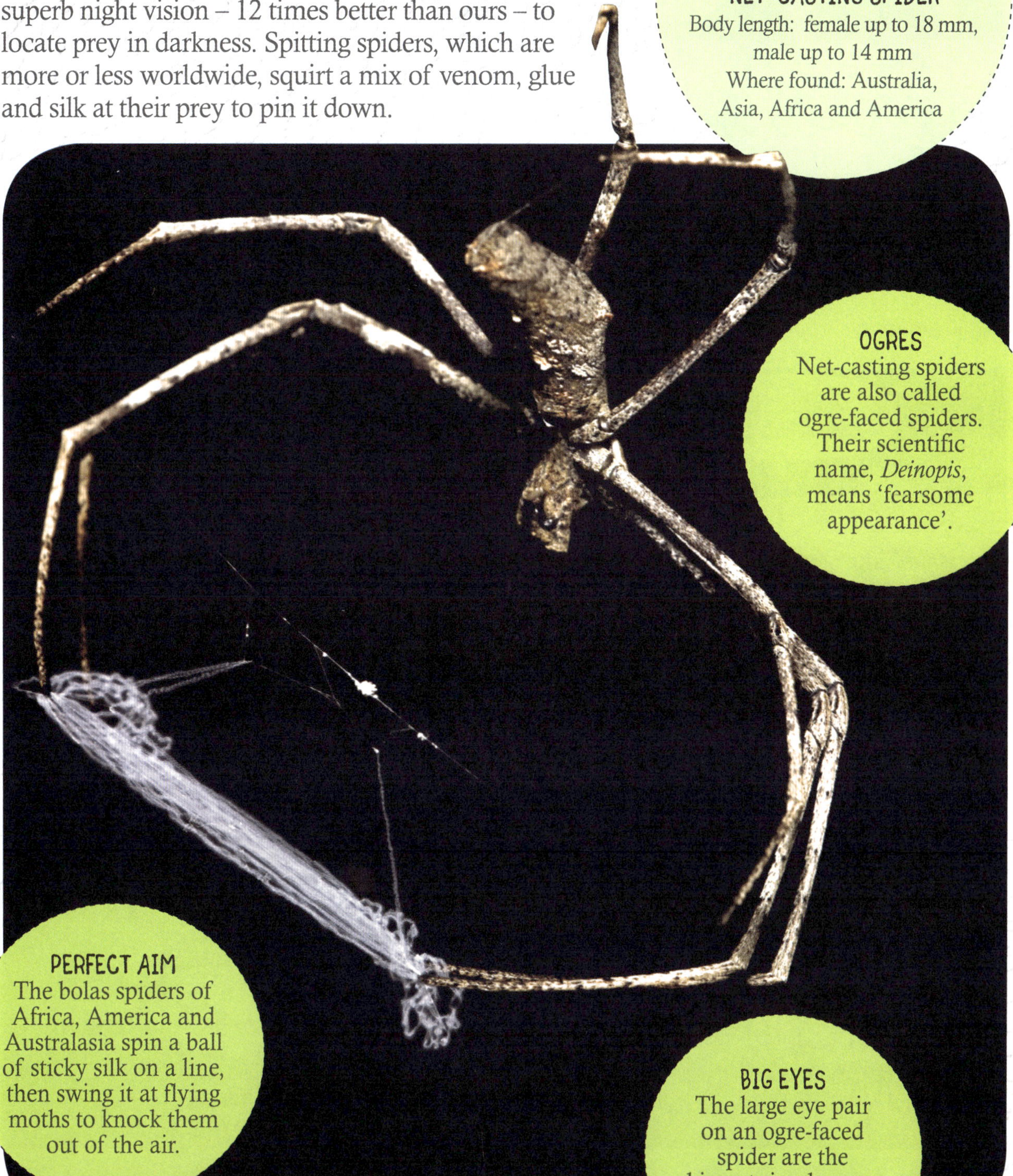

OGRES

Net-casting spiders are also called ogre-faced spiders. Their scientific name, *Deinopis*, means 'fearsome appearance'.

PERFECT AIM

The bolas spiders of Africa, America and Australasia spin a ball of sticky silk on a line, then swing it at flying moths to knock them out of the air.

BIG EYES

The large eye pair on an ogre-faced spider are the biggest simple eyes, relative to body size, of any arthropod.

Fishing

With a film of air around its abdomen, enabling it to breathe, the diving-bell spider spins a canopy of silk underwater, anchoring it to plant stems.

When the spider dives, a 'coat' of air clings to the hairs on its abdomen. The spider hauls its extra-buoyant body down with the help of silken lines.

The air bubble is held underwater in its silken canopy. Once it reaches a certain volume, the bubble 'fills' itself without further effort from the spider, because oxygen naturally filters into it from the water. This 'artificial gill' enables the spider to live underwater like a fish – and to hunt fish!

Raft spiders have a coat of short, velvety hairs that repel water and help them float. Their sensitive feet detect the vibrations of moving prey, such as insects or small fish, which they catch at or below the surface. They can briefly dive, too, trapping a film of air around the abdomen.

WATERY HOME

How does an air-breathing creature spend its whole life underwater? The diving-bell spider, which lives in ponds and rivers, spins itself a silken dome beneath the surface, and fills it with air gathered from above. This oxygen tent becomes a home for the spider, which clambers out to hunt fish and other aquatic life. Raft spiders, too, can hunt on or below the surface, thanks to their amazing 'walk on water' skills.

DIVING-BELL SPIDER
Body length: 8–18 mm
Where found:
Europe and Asia

UNDERWATER EGGS
Diving-bell spiders even lay eggs underwater. A few days after hatching, the spiderlings leave the nest to spin their own tiny diving bells.

BIG SPIDERS!
Raft spiders can be big, with leg spans as wide as the palm of your hand. Some have been known to catch goldfish!

Camouflage

Matching its background perfectly, a crab spider is more or less invisible to its prey: insects that visit plants to collect nectar and pollen.

Like a commando in camouflage gear, the lichen spider is coloured and patterned just like a lichen-covered tree trunk.

Look very carefully near the tideline on American beaches and you may spot the seashore wolf spider – if you can see through its disguise.

The bird dropping spider – disguised as a splotch of poo – sneakily gives off a scent that moths find delicious. They visit... but don't leave!

Some of the most amazing camouflage is seen in tree-dwelling spiders, helping to hide them from birds during the day. Left: the wrap-around spider is named after the way it flattens itself against a branch. Centre/right: at rest, twig spiders look just like stumpy little nubs of wood on a branch.

CLEVER CONCEALMENT

Spiders can be surprisingly hard to spot! Since most of them are sit-and-wait hunters, they have evolved very good camouflage. This not only conceals them from prey, improving their chances of catching a meal, but also helps hide them from predators such as birds. Spider camouflage includes amazing colours and patterns, and unusual body shapes, along with cryptic (hiding) behaviour.

GOLDENROD CRAB SPIDER

Body length: female 6–9 mm, male 3–4 mm
Where found: all around northern hemisphere

BLENDING IN

Usually found on yellow or white flowers, the goldenrod crab spider can change colour to suit its background.

SLOW CHANGE

It takes the spider about six days to change from yellow to white. Changing back takes four times longer, because the spider has to make new yellow pigment.

Mimicry

This *Myrmecium* spider has the usual eight legs, but by waving its long front legs in the air just like antennae, it tricks ants into thinking it's one of them.

A ladybird's bright colours signal to birds that it is foul to eat. So it's no surprise that some spiders – like this *Paraplectana* – mimic the ladybird for protection.

Mimicry can go both ways. On the right is *Coccorchestes*, a jumping spider that mimics a bad-tasting weevil… while on the left is *Agelasta*, a longhorn beetle that mimics a crab spider! It's not clear why – but being a copycat must help the beetle in some way.

Ero cambridgei (left) is a pirate spider that preys on other spiders. Here, it taps the web of a female *Metellina segmentata* in a particular rhythm, to mimic the courtship signals of her mate. This will trick her into coming closer… right into *Ero*'s ambush.

SPIDER TRICKERY

Many spiders have evolved to mimic (look, behave or even smell like) other creatures – such as ants, beetles or even other spiders. This trickery can enable the spider to get close to prey without raising the alarm. It can also give protection from predators, such as birds or spider-hunting wasps. For example, by looking like a bad-tasting beetle, a spider is less likely to be eaten.

ANT-MIMIC SPIDER

Body length: female 6–7 mm, male 5.5–8 mm

Where found: Southeast Asia and Southern China

ENEMIES

Stinging and biting ants are dangerous to spiders – especially when there's a 'gang' of them – so by mimicking an ant, a spider can evade attack.

DISCOVERY

It was English naturalist Henry Walter Bates (1825–1892) who worked out that animals gained protection by mimicking others. He realised this from studying butterflies in the Amazon.

REPRODUCTION

Often the female is much bigger than the male. Just look at this pair of *Nephila* orb weavers. She, after all, will have the job of looking after the kids.

The male *Anyphaena* buzzing spider attracts the attention of a female by vibrating his abdomen noisily. It's a bit like ringing a bell!

To avoid being eaten, the male nursery web spider (left) may soothe a female (right) by offering her a gift of a chewed-up prey animal.

Some spiders wave their legs at each other in a complex 'sign language', to make sure they are ready to mate with one another.

Egg sacs may be buried, hidden, abandoned or guarded carefully, depending on species. Here, a raft spider carries her sac beneath her body, while a bird dropping spider has 'parked' hers on a twig.

FEMALES AND YOUNG

A female spider lays eggs, which hatch into spiderlings. These babies don't have the larval or pupal stages seen in most insects; they are true tiny spiders, and before long they are catching their own prey. Sounds easy? First, spiders need to mate, and it can be very dangerous for a male spider to approach a big, hungry female: she may decide to eat him!

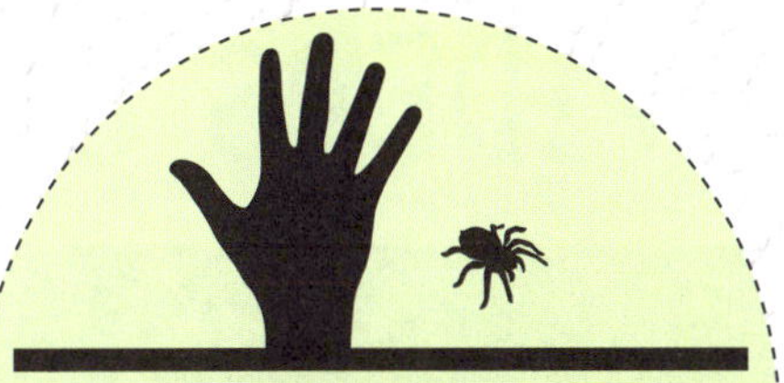

WOLF SPIDER
Body length: 10–35 mm
Where found: worldwide except Antarctica

HITCH-HIKERS
The female wolf spider carries her egg sac on her abdomen. When the spiderlings hatch, they climb up onto her back and hitch a ride.

SILKEN TENT
The female nursery web spider makes a 'tent' of silk on a plant in which her babies can grow safely, while she stands guard outside.

SPIDERS & PEOPLE

The name "tarantula" originally refers to the spider, *Lycosa tarentula*. Long ago, to cure its bite, peasants near Taranto in Italy performed a dance, the 'tarantella'.

In the 19th century, the black widow *Latrodectus mactans* – North America's most venomous spider – often built its web in outside toilets.

Carefully using a pipette, experts 'milk' venom from captive Sydney funnel-web spiders to make medications that can be used to treat spider bites.

Big hairy spiders, like this Mexican red-knee, make popular pets, but overcollection is endangering their populations in the wild.

A single spider can eat about 2,000 insects – flies, mosquitoes, aphids and so on – in a year. That's why many people put up with spiders in the home and do not squash them. So next time a spider builds its web in your house, give it room, and take a closer look...

THE DANGEROUS FEW

The strength of their venom, which can stop prey dead, unfortunately makes a small number of spiders a serious danger to humans. In most species, however, the fangs are just too small to puncture our skin. Also, modern antivenins (medicines) mean that bites are almost never fatal. Nevertheless, you should always treat spiders with respect. Better still, think of them as friends, as they help rid our homes of pests such as flies.

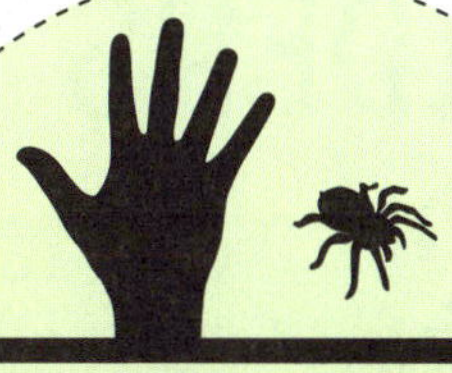

SYDNEY FUNNEL-WEB SPIDER
Body length: 15–45 mm
Where found:
New South Wales, Australia

BEWARE!
About 12 cm across, the Brazilian wandering spider is especially dangerous. It is highly aggressive and injects up to 8 mg of venom – enough to kill about 300 mice.

STRONG BITE
The Sydney funnel-web spider attacks again and again when disturbed. It bites so hard that its fangs can puncture fingernails.

Spiders & Their Enemies

The *Pepsis* tarantula hawk wasp is a huge wasp, up to 50 mm long, with a stinger up to 7 mm long. It specialises in hunting spiders.

Having paralysed a spider with its sting, *Pepsis* drags it back to its burrow in the ground. There, it lays a single egg on the spider and covers it over with soil.

When the wasp larva hatches, it eats the spider – beginning with the non-vital organs so that the spider is kept alive for as long as possible.

Mason and potter wasps also gather paralysed spiders for their larvae, storing them in cells built of mud.

The Australian white-tailed spider (right) hunts at night... for other spiders. Its favourite prey is *Badumna*, a house spider.

With so many enemies, it's no wonder spiders are shy. Jumping spiders, for instance, often hide in a curled-up leaf by day.

A TOUGH LIFE

Being soft and plump, spiders make a nourishing snack for predators, ranging from birds and lizards to insects and even other spiders. Most fearsome of all are the wasps that sting spiders to paralyse them, then store them in a 'living larder' to feed to their larvae. There are flies, too, that burrow into a spider and lay eggs; the fly larvae later eat the living spider. The eight-legged life is tough!

BLACK-BACKED KINGFISHER
Body length: 13 cm
Where found: India and Southeast Asia

SPIDER HUNTERS
Birds snatch spiders from their webs. They also use the silk as a soft lining for their nests. That's one reason why spiders like to hide during the day.

2D VS 3D
It's thought that, about 130 million years ago, spiders evolved 3D tangle webs from 2D orb webs. The more complex 3D webs were a better defence against wasps.

The Other Arachnids

Anything that runs around on eight legs, like the spiders, is classed as an arachnid. There are about 100,000 named species in the class Arachnida and they include ticks, mites, harvestmen, scorpions, solifuges and whip scorpions, among others. As a group they are found all over the world, even in the oceans – and if you look closely, you can see some of them in your backyard.

Ticks are parasites, living on the skin of a host (such as you, or your dog, or a bird). Some are tiny; the biggest are nearly 2.5 cm long. To find a host, many ticks wait with their legs outstretched at the tip of a grass stem. If you brush past them, they jump on. Then they bite into the skin and suck blood, swelling up as they do so. After feeding, they drop off and lay eggs in the ground.

Mites are found in every imaginable land habitat. Worldwide there are more than 50,000 species, feeding on dead plants, dead skin and hair, fresh blood… yuck! Most are too small to see – which is a good thing, because there could be hundreds of dust mites in your bed, right now. They might make you asthmatic or itchy, but otherwise they're harmless.

Harvestmen look a bit like spindly-legged spiders, but, unlike a spider, the two body sections are fused into one. Also, they have just one pair of eyes. They eat almost anything, from dung to plant material to animals alive or dead; unlike spiders, they can digest solid food. You might see a harvestman in sand dunes or heathland, walking jerkily on its thin legs.

Scorpions are found worldwide, with some 1,750 species. Most spend the day under a rock, hunting at night. They use the sharp claws on their palps to tear food apart. All use venom for catching prey, such as insects or mice, and for defence. The venom is injected from the telson, the final tail segment, which can be arched over during attack. Only 25 or so species can kill a human.

Solifuges, also known as camel spiders, sun spiders or wind scorpions, live in dry parts of the world. They hunt anything from beetles to lizards and rodents. Typically they have long, pointed chelicerae (which can make a chattering sound), and very long palps that resemble a fifth pair of legs. At up to 15 cm long, solifuges look alarming, but pose little threat to humans.

Whip scorpions include two groups. Amblypygids (above) live in warm places. They have a flattish body and eight long legs, although the first pair are used as antennae, not for walking. Vinegaroons look a bit like scorpions, but with a whip-like 'tail'. When disturbed, they can spray a sharp-smelling chemical, hence their name.

Hello, eight-legged friends!

Six spider facts

The female desert spider *Stegodyphus lineatus* rears just one lot of spiderlings in her life, and literally dies for her babies. Her digestive juices soften the food in her stomach, which she then vomits up. Once the spiderlings have eaten that up, they devour their mother, leaving just a dry empty husk. Then they leave the nest.

One of the most venomous North American spiders is the brown recluse or fiddleback, named after the violin-shaped markings on its abdomen. Its bite can kill young children, but luckily the spider is shy and attacks only reluctantly.

The six-eyed sand spider *Sicarius hahnii*, which lives in desert regions of Southern Africa, can go 12 months without food or water.

Really big mygalomorph spiders can kill and eat snakes – even 45-cm rattlesnakes. They typically go for the snake behind the head, inflicting a fatal bite.

In parts of Cambodia, especially the town of Skuon, locals serve up crispy-fried spiders, each about as big as your hand. The taste, apparently, is halfway between chicken and fish!

The spiders in one family, the Uloboridae, have no venom fangs. Instead, they kill prey by wrapping it in very fuzzy silk – sometimes hundreds of metres of it – which eventually crushes the captive. Then they vomit digestive juices over the victim to soften it into an edible 'soup'.

Glossary

abdomen the hind part of a spider's body.

antenna (plural: antennae) one of the two slender, movable sensory organs on an insect's head, providing touch, taste and smell. In some insect-mimicking spiders, the front legs look very like antennae.

Arachnida the class containing the spiders and other eight-legged invertebrates. Members of the Arachnida are called arachnids.

cephalothorax the 'head' part of a spider's body, made up of the cephalon (head) and thorax (mid part).

chelicera (plural: chelicerae) one of the pair of appendages on either side of the spider's mouth. Each chelicera is tipped with a hollow fang.

cribellum an organ beneath the abdomen, used for spinning silk.

fang the hollow, pointed part of a chelicera, used for biting into prey and injecting venom.

mimicry copying the appearance, behaviour or some other features of another animal.

mygalomorph a primitive kind of spider in which fangs operate up and down, not in a pincer action. Mygalomorphs include the big hairy spiders that most people refer to as 'tarantulas'.

palp one of the pair of organs attached to the spider's head, which is used by a male to transfer sperm to a female. It can also be used as a feeler.

spinneret one of the nozzle-like organs at the tip of a spider's abdomen, used for squeezing out lines of silk.

tarantula the name belonging properly to the wolf spider *Lycosa tarentula*, but also used for any large hairy spider in the United States, or the huntsman in Australia.

telson the tip of a scorpion's tail, holding the venom gland and stinger.

venom poison that is injected (from a fang or a stinger) into prey.

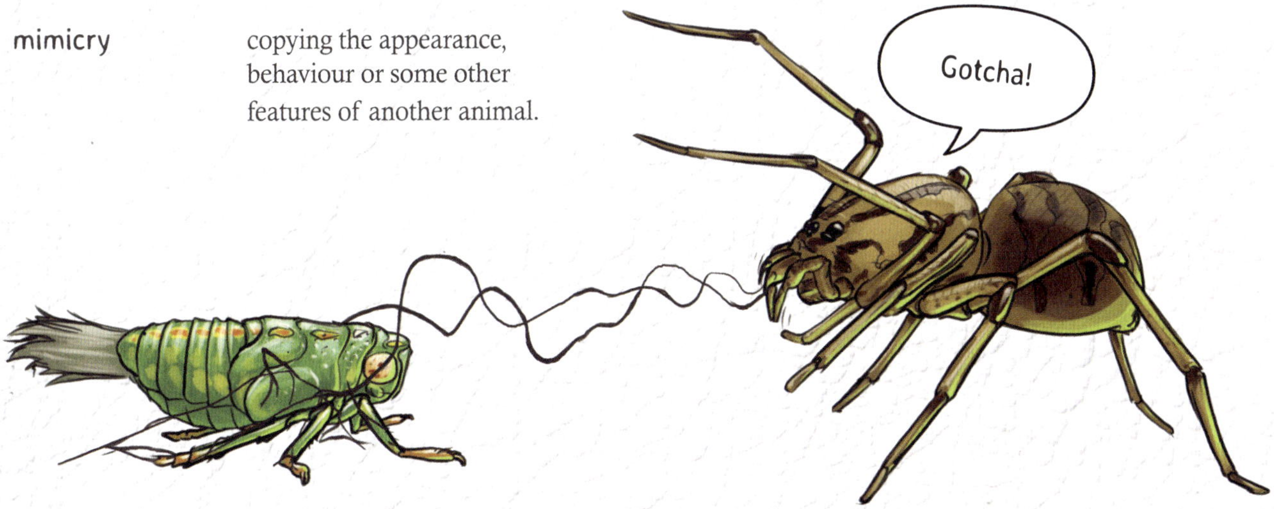

INDEX

The Author

British-born Matt Turner graduated from Loughborough College of Art in the 1980s, since which he has worked as a picture researcher, editor and writer. He has authored books on diverse topics including natural history, earth sciences and railways, as well as hundreds of articles for encyclopedias and partworks, covering everything from elephants to abstract art. He and his family currently live near Auckland, New Zealand, where he volunteers for the local Coastguard unit and dabbles in art and craft.

The Artist

Born in Medellín, Colombia, Santiago Calle is an illustrator and animator trained at Edinburgh College of Art in the UK. He began his career as a teacher, which led him to deepen his studies in sequential art. Santiago founded his art studio Liberum Donum in Bogotá in 2006, partnering with his brother Juan. Since then, they have dedicated themselves to producing concept art, illustration, comic strip art and animation.

Picture Credits (abbreviations: t = top; b = bottom; c = centre; l = left; r = right)
© www.shutterstock.com:

1 cb, 2 cl, 4 c, 6 tl, 7 tl, 7 tr, 7 bl, 9 c, 13 c, 15 c, 17 c, 19 c, 21 c, 23 c, 25 c, 27 c, 28 tr, 28 cl, 28 br, 29 tl, 29 cr, 29 bl, 32 cr.

11 c © Bryce MQuillan NZ

索引

作者简介

马特·特纳出生于英国，20世纪80年代毕业于拉夫伯勒大学艺术学院，毕业后一直担任图片研究员、编辑和作者。他的书题材广泛，涉及博物学、地球科学和铁路等，并为百科全书和分册出版的丛书写过数百篇文章，从大象到抽象艺术无所不包。他现在和家人住在新西兰奥克兰附近，他还是当地海岸警卫队的志愿者，平时也涉猎工艺品的制作。

绘者简介

圣地亚哥·卡列出生于哥伦比亚的麦德林，是一位插画师和动画师，曾在英国爱丁堡艺术学院接受过培训。他的第一份职业是教师，教学促使他在连续性艺术领域继续深造。2006年，他和兄弟胡安在波哥大合伙创立了一家名为“自由德南”的艺术工作室，自此两人便致力于概念艺术、插画、连环画和动画的创作。

Picture Credits (abbreviations: t = top; b = bottom; c = centre; l = left; r = right)
© www.shutterstock.com:

1 cb, 2 cl, 4 c, 6 tl, 7 tl, 7 tr, 7 bl, 9 c, 13 c, 15 c, 17 c, 19 c, 21 c, 23 c, 25 c, 27 c, 28 tr, 28 cl, 28 br, 29 tl, 29 cr, 29 bl, 32 cr.

11 c © Bryce MQuillan NZ

术语表

螯牙　螯肢内中空、尖利的部分，用来咬住猎物，注射毒液。

螯肢　蜘蛛口部两边的附肢（共一对），尖端有空心的螯牙。

触角　昆虫头部细长可移动的感觉器官（通常为两根），具有触觉、味觉和嗅觉功能。有些蜘蛛伪装成昆虫的模样，它们的前足看起来就像是一对触角。

触肢　附着于蜘蛛头部的附肢（共一对），雄蛛用附肢将精子传送给雌蛛。触肢也可发挥类似昆虫触角的功能。

毒液　通过牙或螯刺注入猎物体内的有毒液体。

纺器　蜘蛛腹部末端的小管状器官，用来排出蛛丝。

腹部　蜘蛛身体分为头胸部和腹部两部分，腹部是后半部分。

猛蛛　一种较为原始的蜘蛛种类，其螯牙上下活动，而不是像钳子一样侧向开合。猛蛛中有体形较大、浑身长满绒毛的蜘蛛，大多数人将其称为“塔兰托毒蛛”。

拟态　动物在外观、行为等特征上模拟其他动物的现象。

筛器　蜘蛛腹部下面的器官，用于排出蛛丝。

塔兰托毒蛛　这个名字严格意义上指的是塔兰托狼蛛。但在美国，体形较大、毛茸茸的蜘蛛都被称为塔兰托毒蛛。在澳大利亚，猎蛛也被称为塔兰托毒蛛。

头胸部　蜘蛛身体的前半部分，包括连在一起的头部和胸部。

尾节　蝎子尾巴的末端，内有毒液腺和螯刺。

蛛形纲　蛛形纲包括蜘蛛和其他八条腿的无脊椎动物，其中的成员被称为蛛形纲动物。

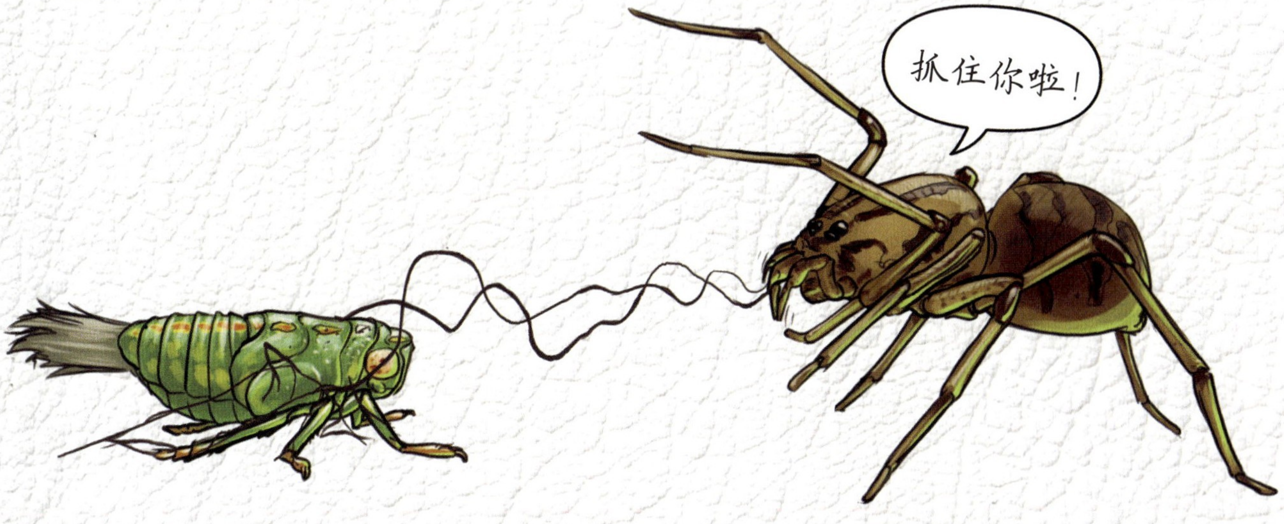

蜘蛛奇闻

雌性沙漠穹蛛一生只生育一次，毫不夸张地说，它会“为孩子献出生命”。它用消化液将胃中的食物软化后再吐出来。一旦幼蛛吃完这些食物，就会吞食它们的母亲，最终只剩下一个干枯的蜘蛛空壳。之后，幼蛛离开巢穴。

在北美，有一种毒性极强的褐隐士蛛，因其腹部有小提琴形状的斑纹，也被称为“小提琴蛛”。它叮咬婴幼儿后，可能会导致婴幼儿死亡。好在这种蜘蛛性情胆怯，只有在迫不得已的情况下才会出击。

在没有食物和水的条件下，生活在非洲南部沙漠地区的六眼沙蜘蛛可以存活12个月。

体形巨大的猛蛛能杀死一条蛇并将其吃掉，就连长达45厘米的响尾蛇都不在话下。它们通常从蛇头后部袭击，造成致命的咬伤。

在柬埔寨的一些地区，尤其是素昆镇，当地人会把脆炸蜘蛛端上饭桌，每只蜘蛛大概有手掌那么大。至于味道，据说介于炸鸡和炸鱼之间。

妩蛛科蜘蛛的螯牙无毒，因此它们用毛茸茸的蛛丝作为替代武器。妩蛛的蛛丝有时长达数百米，能将猎物裹起来，最终将其挤压至死。然后，妩蛛将消化液吐到猎物身上，将猎物软化成为可以食用的“汤”。

蝎子

蝎子遍布世界各地，约有1,750个品种。大多数蝎子白天藏在岩石下，夜间出来捕食。它们用触肢上的尖爪撕裂猎物。所有的蝎子都用毒液来猎捕昆虫、老鼠等猎物，并进行防御。蝎子从尾节中喷射出毒液，尾节位于尾巴末端，在进攻时可以向上卷起。只有大约25种蝎子能置人于死地。

避日蛛

避日蛛也被称为骆驼蜘蛛、日蛛或风蝎，生活在气候干燥的地区。它们捕食甲虫、蜥蜴、啮齿动物等。避日蛛的螯肢通常又长又尖，能发出“吱吱”的声音，触肢也非常长，就像是第五对足。避日蛛体长可达15厘米，看起来很可怕，但实际上对人类构不成多大的威胁。

鞭蝎

鞭蝎包括两类。第一类是无鞭蝎，如上图所示。它们生活在气候温暖的地区，身体扁平，有四对足，不过第一对足并不是用来走路的，而是发挥触角的功能。第二类是巨鞭蝎，样子有点儿像蝎子，但“尾巴”像一条鞭子。当受到惊扰时，它们能喷射出带有刺鼻气味（浓烈醋味，译注）的化学物质，因此也被称为“喷醋鞭蝎”。

其他蛛形纲动物

像蜘蛛这样用八条腿走路的生物都属于蛛形纲。蛛形纲中已被命名的物种大约有100,000种，包括蜱虫、螨虫、盲蛛、蝎子、避日蛛、鞭蝎等。蛛形纲家族成员遍布世界各地，甚至也出现在海洋里。如果你仔细观察，在你家后院就能发现它们的踪迹。

蜱虫

蜱虫是寄生虫，寄生在宿主（人、狗、鸟等）的皮肤上。有的蜱虫非常小，而最大的蜱虫将近2.5厘米。为了寻找宿主，许多蜱虫在草茎尖上伸着腿等待。如果你恰巧擦身而过，蜱虫就会跳到你身上，叮咬你的皮肤，吸食你的血液。蜱虫吸血后身子会胀大。吸饱后，它们会落在地面上产卵。

螨虫

螨虫出没于你可以想象到的任何一个陆地栖息地。世界范围内有超过50,000种螨虫，它们以死去的植物、死皮、毛发、新鲜的血液等为食。够恶心的吧！大多数螨虫太小，单凭肉眼看不到——这样也好，因为此时此刻你的床上可能有成百上千只尘螨。尘螨可能引发哮喘和瘙痒，不过除此之外，它们也没有其他的危害。

盲蛛

盲蛛看起来有点儿像细腿的蜘蛛，但蜘蛛的身体分为头胸部和腹部两部分，而盲蛛的两部分是合二为一的。另外，盲蛛只有一对眼睛。盲蛛几乎无所不吃，动物的粪便、植物、动物的尸体和活体，它们照单全收。盲蛛可以消化固体食物，蜘蛛则不能。在沙丘或荒野中，你或许能发现盲蛛用细长的腿高一脚低一脚地行走。

“蛛生”不易

蜘蛛身体柔软、肉乎乎的，是捕食者的营养点心。蜘蛛的天敌有鸟类、蜥蜴、昆虫，甚至还有自己的同类。其中最可怕的就数胡蜂，它们能蜇刺蜘蛛，将蜘蛛麻醉后贮存于“生鲜储藏柜”里，以喂养幼虫。还有一些蝇类在蜘蛛身上挖洞，将卵产在蜘蛛体内，生出的蛆虫继而将蜘蛛的活体蚕食殆尽。这种八条腿的生物，生活真是太艰辛啦！

黑背翠鸟

体长：13厘米

分布区域：印度、东南亚

蜘蛛猎手

鸟类从蛛网中抓走蜘蛛，甚至还用蛛丝作为鸟巢的软衬。这也是蜘蛛喜欢在大白天“躲猫猫”的原因之一。

二维网和三维网

人们通常认为大约1.3亿年前，蜘蛛改良了织网技术，从织造二维的圆网改为织造三维的立体不规则网。三维网更加复杂，能够更好地防御胡蜂。

蜘蛛VS天敌

蛛蜂属鳖甲蜂是一种巨型胡蜂，体长可达50毫米，螫刺长达7毫米，专门捕食蜘蛛。

鳖甲蜂将蜇咬后陷入麻痹状态的蜘蛛拖回地洞，并在蜘蛛的身体上产下一个卵，然后用土把蜘蛛盖起来。

鳖甲蜂的幼虫孵化后，蜘蛛便成了它的美餐。它从不那么重要的器官开始吃起，好让蜘蛛活得尽可能久一点。

蜾蠃蜂也会将蜘蛛蜇刺麻醉后，储藏在用泥土砌成的巢室里，作为幼虫的食物。

上图右侧是澳大利亚的白尾蜘蛛，它在夜间捕食其他种类的蜘蛛。它最爱捕食社蛛这种家蜘蛛。

蜘蛛的天敌众多，难怪蜘蛛性情胆怯。比如，跳蛛白天会躲藏在卷曲的叶子里。

少数危险分子

蜘蛛的毒液可以让猎物一命呜呼，这样的毒性使一小部分蜘蛛对人类也构成了严重的威胁，实为不幸。不过，大多数蜘蛛的螯牙太小，无法刺穿我们的皮肤。而且，有了现代的抗毒素药物，蜘蛛咬伤一般是不会致死的。尽管如此，你还是应该时刻小心蜘蛛。更明智的做法，是把蜘蛛当成朋友，因为它们能帮助我们消灭家中的苍蝇等害虫。

悉尼漏斗网蜘蛛

体长：15—45毫米

分布区域：澳大利亚新南威尔士州

当心！

巴西游走蛛的身体宽约12厘米，危险系数极高。它生性极其凶猛，能向猎物体内注入多达8毫克的毒液，足以杀死300只老鼠。

厉害的螯牙

悉尼漏斗网蜘蛛在受到惊扰时会一次又一次地发动进攻。它下口极狠，那副螯牙足以刺穿人的指甲。

蜘蛛和人

“塔兰托毒蛛”最初仅指“塔兰托狼蛛”。很久以前，意大利塔兰托附近的农民被这种蜘蛛咬伤后，会跳塔兰台拉舞来疗伤。

19世纪，北美洲毒性最强的黑寡妇蜘蛛经常在户外的厕所里织网。

专家小心翼翼地用吸量管吸取捕获的悉尼漏斗网蜘蛛的毒液，用来制作治疗蜘蛛咬伤的药。

体形较大、长满绒毛的蜘蛛是很受欢迎的宠物，比如墨西哥红膝鸟蛛。然而，由于过度捕捉，这些蜘蛛的野生数量已经遭到了威胁。

一只蜘蛛一年可以吃掉2,000只昆虫，这其中包括苍蝇、蚊子、蚜虫等。这就是为什么许多人可以容忍蜘蛛与自己同在一个屋檐下，而不把它们拍死。所以，下次蜘蛛在你家织网的时候，给它些空间吧，顺便近距离观察一下……

雌蛛和幼蛛

雌蛛产卵，卵孵化成幼蛛。幼蛛不会经历大多数昆虫所经历的幼虫或蛹的阶段，它们已经是真正的小小蜘蛛了。不久以后，幼蛛就可以自己捕食猎物了。听起来很简单吧？可是，首先蜘蛛需要交配。对于雄蛛而言，接近一只饥肠辘辘的大个头雌蛛可谓“九死一生”——雌蛛可能会决定吃掉它！

狼蛛

体长：10—35毫米

分布区域：除南极洲外世界各地

搭便车

狼蛛妈妈将卵囊挂在肚子下面。幼蛛孵化出来后，会爬到妈妈的背上，搭顺风车。

蛛丝帐篷

盗蛛妈妈会在植物上用蛛丝织出一个“帐篷”，并在一旁守护，让盗蛛宝宝们在帐篷里安全地成长。

生儿育女

通常来说，雌蛛比雄蛛的块头要大得多。看看这对络新妇属金丝蛛你就明白了。毕竟，雌蛛还要照料儿女呢。

近管蛛属雄蛛通过腹部的振动发出嗡嗡声，来引起雌蛛的注意，整个过程有点像按铃。

雄盗蛛（上图左）会将捕获的猎物嚼碎后作为礼物送给雌蛛（上图右），以抚慰雌蛛，请它口下留情，饶自己性命。

有些蜘蛛会用复杂的“手语”交流——雌蛛和雄蛛朝彼此挥舞足部，表示它们已经准备好要与对方进行交配了。

不同种类的蜘蛛对卵囊有不同的处理方式——埋起来、藏起来、遗弃或者小心守护。上图中，一只水涯狡蛛把卵囊带在腹部下面，而鸟粪蛛则把卵囊挂在了树枝上。

蜘蛛的小把戏

许多蜘蛛在进化中获得了拟态的本领，能在外观、行为乃至体味上模拟蚂蚁、甲虫等其他生物，甚至能模拟其他种类的蜘蛛。通过拟态，蜘蛛可以神不知鬼不觉地靠近猎物，还可以避免受到鸟类、食蛛蜂等捕食者的攻击。例如，蜘蛛通过模仿一只味道糟糕的甲虫，可以大大降低被捕食的可能性。

蚁蛛

体长：雌蛛6—7毫米，
雄蛛5.5—8毫米
分布区域：东南亚、
中国南部

蜘蛛的天敌

对蜘蛛而言，被蚂蚁叮咬很危险。如果蚂蚁成群结队行动，那就更是大事不妙了。于是，蜘蛛通过模仿蚂蚁，来逃避蚂蚁的攻击。

科学家的发现

英国博物学家亨利·沃尔特·贝茨（1825—1892）在亚马孙河研究蝴蝶时意识到，动物的拟态行为是为了保护自己，增加生存机会。

拟态

上图中的蚁蛛和其他蜘蛛一样，也长了八条腿。不过，它们会将最前面的一对足举到空中，像触角一样来回舞动，让蚂蚁误把它们当成自己的同类。

瓢虫利用鲜艳的警戒色警告鸟类：“我可是很难吃哦。”所以，瓢蛛等蜘蛛模仿瓢虫来保护自己，也就不足为奇了。

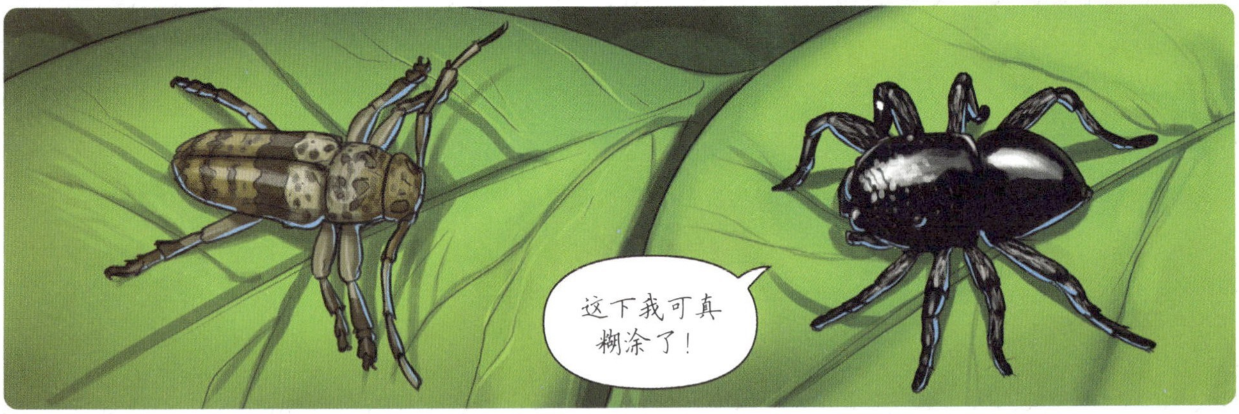

蜘蛛能模仿其他动物，其他动物也能模仿蜘蛛。瞧，上图中右边的谷跳蛛属蜘蛛模仿一条味道糟糕的象鼻虫，而左边这只天牛却模仿了一只蟹蛛。人们还不清楚天牛模仿蜘蛛的原因，不过肯定是对自己有好处的。

上图左边是一只坎氏突腹蛛，它是一种盗蛛，以其他蜘蛛为食。图片右边的蜘蛛名叫节后鳞蛛。坎氏蛛正在以一种特殊的节奏敲击这只雌性节后鳞蛛的网，这是为了模拟雄蛛的求偶信号，哄骗雌蛛走进它设下的伏击圈。

巧妙的隐藏术

辨认蜘蛛可是一件出奇困难的事情！大多数蜘蛛都采用“守株待兔”的方式捕食，因此进化出了高超的伪装术。这一方面是为了避免被猎物发现，增加捕食成功的可能性；另一方面也是为了躲避鸟类等天敌。蜘蛛的伪装术体现在它们不可思议的色彩和图案、不同寻常的体形，以及隐秘的隐藏行为中。

秋麒麟蟹蛛

体长：雌蛛6—9毫米，
雄蛛3—4毫米
分布区域：北半球

融入环境

秋麒麟蟹蛛常见于黄色或白色的花朵上，能根据周围环境改变体色。

缓慢的变色过程

秋麒麟蟹蛛从黄色变为白色大约需要六天，而从白色变回黄色则需要花费四倍的时间，因为秋麒麟蟹蛛需要在体内生成新的黄色素。

伪装术

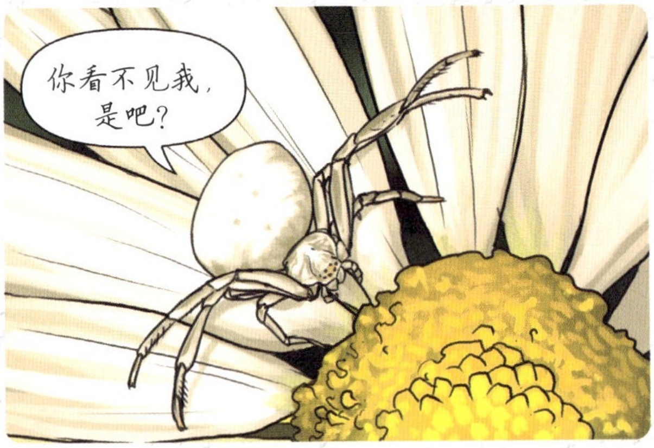

蟹蛛以采集花蜜和花粉的昆虫为食。它可以完美地融入周围环境中，很难被猎物觉察出来。

地衣鬼蛛就像是一名伪装突击队员，它的体色和斑纹恰如一根长着地衣的树干。

海滨狼蛛栖息在美国海滩的涨潮线附近，如果你仔细观察，兴许能识破伪装，侦查到目标。

鸟粪蛛将自己伪装成一坨便便，悄悄地散发出一种蛾子喜欢的气味。蛾子一旦被吸引过去，就再也走不了了！

树栖蜘蛛是伪装高手，它们凭着高超的伪装技艺可以在白天躲过鸟类的攻击。上图左：环扁园蛛将身子摊平紧紧地贴在树枝上，它也因此得名。上图中和上图右：正在休息的蚓腹蛛看起来就像树枝上短短粗粗的小结节。

水下的家

一种需要呼吸空气的生物，却一辈子生活在水下，这是怎样做到的呢？生活在池塘和河流里的潜水钟蜘蛛为自己在水下织了一个丝质的半球形房子，并填充入从水上收集的空气。这个氧气帐篷就成了潜水钟蜘蛛的家，需要捕食鱼类和其他水生生物的时候，它们就从帐篷里爬出来。而水涯狡蛛身怀“水上走”绝技，所以无论是在水上还是水下，它们都能够捕食。

潜水钟蜘蛛

体长：8—18毫米

分布区域：欧洲、亚洲

水下产卵

潜水钟蜘蛛连产卵也是在水下进行。幼蛛孵化出来几天后就会离开妈妈，织造自己的小潜水钟。

巨型蜘蛛

水涯狡蛛体形巨大，腿的跨度甚至可以和人的手掌一样宽。有些水涯狡蛛甚至可以捕食金鱼！

捕鱼

这位置不错。
视野开阔，还有美味的邻居……

这种时候，我总是要感谢重力的存在。

潜水钟蜘蛛腹部周围有一层空气膜，使它可以在水下呼吸。它在水下用蛛丝织出一个天篷，固定在植物的茎上。

当潜水钟蜘蛛潜入水下时，会有一层空气附着到它腹部的毛上。虽然潜水钟蜘蛛受到的浮力非常大，但它可以利用蛛丝拖曳将身子沉入水下。

那是什么？
是一条怪鱼吗？
你想得美。

潜水钟蜘蛛将气泡保存在水下的蛛丝天篷里。一旦气泡变得足够大，就会自动充气，而不再需要潜水钟蜘蛛费力，因为氧气可以自动从水中扩散到气泡里。这种“蛛造腮”使潜水钟蜘蛛像鱼一样生活在水里，并且还能捕食鱼类！

谁需要钓鱼竿？

水涯狡蛛身上有一层天鹅绒般的短毛，这层短毛具有防水功能，能帮助它们浮在水面上。水涯狡蛛的脚十分敏感，可以感知到昆虫、小鱼等猎物在运动时产生的振动，无论猎物是潜在水下，还是逃到了水面上，都能被它们擒获。它们也能在腹部周围贮存一层空气，进行短时间潜水。

伏击捕食者

撒网蛛喜欢温暖的环境，从南美洲国家到马来西亚再到澳大利亚，均有撒网蛛栖身。它们织出的网很小巧，如同一张小渔网。撒网蛛设下埋伏，坐等猎物路过，然后将网撒出去。它们具有超强的夜视能力，是人类的12倍，可以在黑暗中准确地定位猎物。喷液蛛几乎分布在世界各地，它向猎物喷射的是由毒液、黏性物质和蛛丝组成的混合物，这一喷便能让猎物动弹不得。

撒网蛛

体长：雌蛛长达18毫米，
雄蛛长达14毫米。
分布区域：澳大利亚、
亚洲、非洲和美洲

妖面蛛

撒网蛛又称为“妖面蛛”，拉丁文学名是“Deinopis”，意思是“可怕的外表”。

完美命中

流星锤蜘蛛分布在非洲、美洲和大洋洲，它们在蛛丝末端织出具有黏性的丝球（外形就像古代的兵器流星锤，译注），然后冲着飞蛾挥动丝球，将飞蛾从空中粘下来。

巨眼蜘蛛

撒网蛛的“妖面”上有一对大眼睛，相对于它的体长来说，它的单眼在节肢动物中是最大的。

撒网和喷液

撒网蛛首先织出起支撑作用的“脚手架”结构，接着用极富弹性的蛛丝在后足上织出一张小巧的“渔网”。

现在，撒网蛛倒挂金钩，并用四只前足抓着小“渔网”，时刻准备着向猎物进攻。

一旦甲虫进入撒网蛛的攻击范围，它就会把网拉伸到高达原来大小的十倍，撒向猎物。一股一股的蛛丝又细又卷，是从筛器（参见本书第8页）中排出的，能将猎物彻底缠住。

喷液蛛能够从螯牙中喷出两股带有毒液的蛛丝。与此同时，它把身子左右摇晃，使蛛丝呈现锯齿状，像一张黏糊糊的网把猎物粘住。整个攻击过程大约只有0.03秒，速度非常快，只有通过慢镜头才能看到。

下嘴有毒

几乎所有的蜘蛛都借助毒液来捕食：猛地一口咬下去，猎物大餐就停止了挣扎！当蜘蛛将中空的螯牙刺入猎物体内时，烈性毒液就顺着流进去，麻痹或杀死猎物，于是蜘蛛就可以从容不迫地进食了。毒液会软化猎物的内脏，将其变成“汤汁”，这样蜘蛛就可以吸食啦。当然，做汤之前，还是得先打猎才行。

狼蛛

体长：10—35毫米
分布区域：除南极洲外世界各地

螯牙

柯氏石蛛只吃一种东西。你能猜出来吗？对啦，就是鼠妇。这种蜘蛛的螯牙极为强壮，可以刺穿鼠妇的外骨骼。

毒液

蜘蛛的毒液主要有两种。一种是神经毒素，可以攻击猎物的神经系统，让猎物动弹不得。另一种是细胞毒素，可以溶解猎物的内脏。蜘蛛可能会拥有其中某一种毒液，也可能两者均有。

毒液和捕食

看我“鹤嘴镐”式攻击！

夹取东西对我来说就是小菜一碟。

蜘蛛的螯牙有两种工作方式。第一种方式的代表是猛蛛（直颚亚目），它们的螯牙上下活动，有点儿像一对鹤嘴镐。

第二种方式的代表是被称为“真正”的蜘蛛的原蛛（钳颚亚目），它们的螯牙侧向开合。

抓住你啦！

嗯，我喜欢这些填满馅儿的昆虫卷。

蝗蛸会坐等猎物上钩。当猎物靠近时，它们从洞穴中猛蹿出来，抓住猎物，拖进洞穴。

如果有昆虫无意中闯入圆蛛网，蜘蛛在短短5—10秒内就可以冲上去，咬住猎物，然后用蛛丝将猎物裹起来吃掉。

看，忍者来啦！

跳蛛会像老虎一样向猎物猛扑过去。世界上有超过5,000种跳蛛。跳蛛有八只眼，中间的一对特别大，能直视前方，帮助它们在几厘米范围内精准地判断距离的远近。（这两只眼也使得它们的模样很可爱。）

蛛网陷阱

古老的蜘蛛住在地洞里。直到今天，有些蜘蛛仍然沿袭着这样的生活方式，尤其是比较原始的猛蛛。蠖蛸通常会在洞口加设一个可以活动的盖子，藏起入口处。与蠖蛸不同，漏斗蜘蛛用蛛丝“绊线”将入口圈起来，一旦猎物走近，蛛丝“绊线”便会发出警报。一些生活在地面上的蜘蛛会给蜘蛛网加上保护性的通道。

霍氏六纺蛛

体长：最长可达20毫米

分布区域：新西兰

超级大力士

当拉力达到自身重量的38倍时，加利福尼亚大蠖蛸依然可以借助倒钩和螯牙让洞口活盖保持关闭。此举可是创了纪录的哦！

挖掘大师

蠖蛸可以挖出深达30—40厘米的地洞。它的寿命高达30年。

地洞和活板门

快来看看
我新挖的洞。

穴居蜘蛛用带刺的螯牙在土里刨出一个洞，然后用一层层的泥土、唾液和蛛丝对洞穴壁进行防水处理。

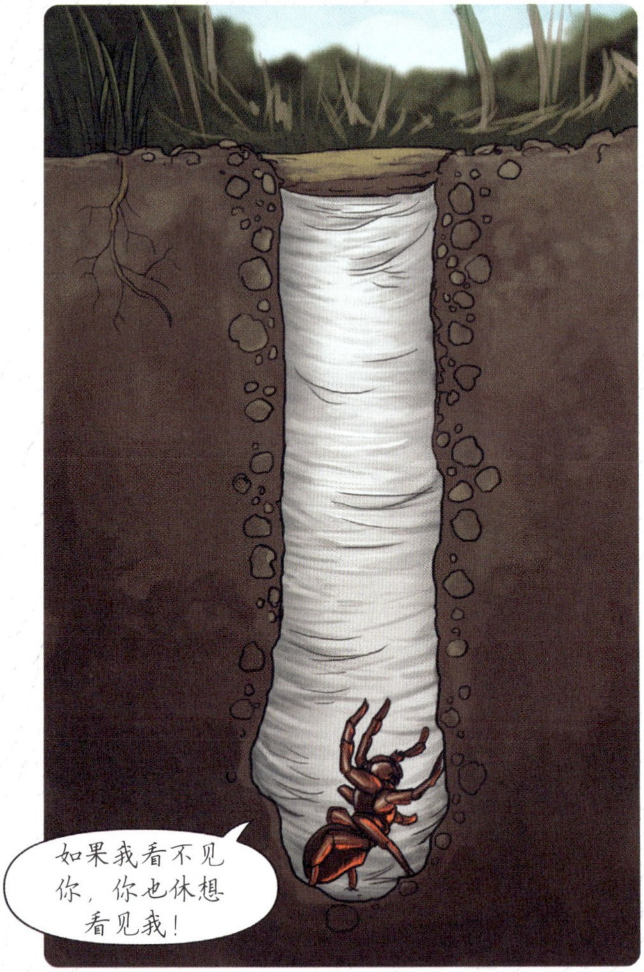
如果我看不见你，你也休想看见我！

“活板门”关闭后，蝗蛸的洞穴就藏在暗处看不见了。不过，食蛛蜂往往能够找到洞穴的入口。如果大水冲来，这扇门也是靠不住的。

啊，
送上门来啦！

蝗蛸在洞穴口用蛛丝、泥和草做了一扇“活板门”，平时藏在活盖下，埋伏起来坐等猎物靠近。

大小刚刚好，
就是我想要的漏斗形！

漏斗蛛科的蜘蛛在地面上织出片状网，网上设有漏斗形的藏匿处。当猎物出现时，蜘蛛会从藏匿处突然冲出来抓住猎物。

知道不？我可不光有颜值。

蝗蛸科盘腹蛛属的蜘蛛腹部末端扁平，上有坚硬的腹盘。在遇到威胁时，盘腹蛛便把自己“塞”进洞穴里，腹部堵在洞口。

纺丝

蛛丝在蜘蛛体内为液态，从纺器中排出体外后变干，形成线状物。蜘蛛可以纺出不同类型的丝：有黏性的用来粘住猎物，没有黏性的用来行走，韧性特别强的用来悬挂自己，如此等等。蜘蛛最为著名的作品就是美丽的螺旋形圆蛛网，不过，许多蜘蛛结的网是更具立体感的“不规则网”，看起来要杂乱一些。

金丝蛛

体长：雌蛛长达50毫米，
雄蛛5—8毫米
分布区域：澳大利亚、
美洲、非洲和亚洲

坚韧的蛛丝

蛛丝的强度是相同粗细的钢丝的五倍。也许有一天，人们可以用蛛丝制造军用防弹背心。

园蛛

园蛛编织一张普通的圆网大概要用长达60米的蛛丝，但一般来说，它一个小时内就能织完。

蛛丝和圆形蛛网

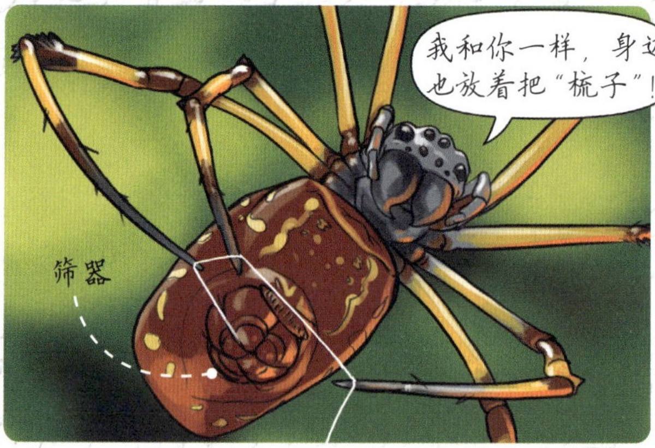
我和你一样，身边
也放着把“梳子”！
筛器

有些种类的蜘蛛长有筛器，筛器上分布着许多吐丝管。蜘蛛从筛器中排出纤细的丝，然后用步足将丝梳理得毛茸茸的，为猎物布下天罗地网。

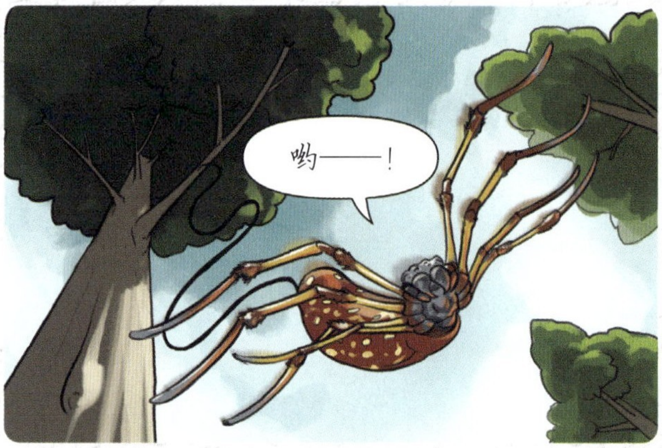

蛛丝富有弹性，且很有韧性。当蜘蛛从树枝之类的歇脚点跳下来或落下来时，经常会织一根“牵引丝”，接住落下的自己。

我还没有彻底
完工呢！

由于蛛丝质地坚韧，早期在新几内亚部落里，人们把蛛丝织在木质环状渔网上，用它到河里捕鱼。

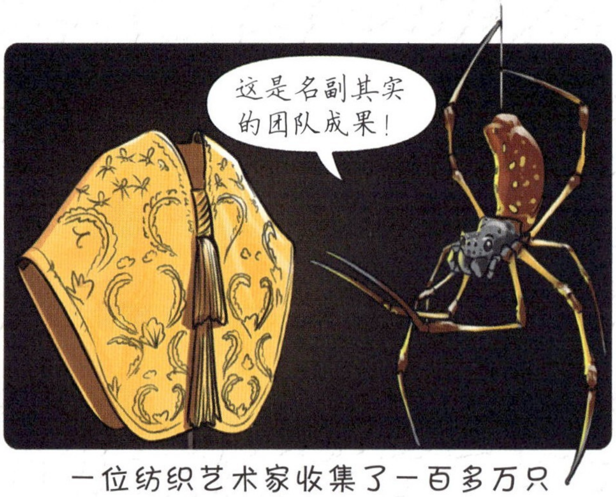
这是名副其实
的团队成果！

一位纺织艺术家收集了一百多万只金丝蛛纺出的金色蛛丝，织成了这件美丽的披肩。

太棒了，我的跑道
已经标在那儿了！
我要
着陆啦！

有些蜘蛛在织网时会织出“之”字形图案，被称为“隐带”。这些吸引眼球的设计目的何在？是为了反射紫外线引诱昆虫自投罗网？还是要使蛛网更显眼，从而防止鸟类靠近？抑或是为了让蜘蛛看起来个头更大？对此，专家也不能确定。

蜘蛛是捕食性动物，有的织网捕食，有的伏击捕食，还有的主动追捕猎物。蜘蛛吃不了固体食物，因此大多数蜘蛛会用毒液杀死或麻痹猎物，将其内脏做成“内脏汤”。有些种类的蜘蛛，毒液毒性特别强，对人有害甚至能致命。但绝大部分蜘蛛不仅对人类无害，还可以帮助我们消灭虫害。据估计，蜘蛛每年捕食的猎物的重量总和超过了全人类的重量总和。

惊人的蜘蛛

提起蜘蛛，很多人心里都会发毛，但蜘蛛也有令人着迷的一面。它们已经在地球上生活了三亿多年，历经漫长的进化，将领地拓展到几乎所有陆地栖息地——深洞、沙漠、雨林，到处都有蜘蛛生息。即便是在我们住的房子里，蜘蛛也不拿自己当外人。全世界除了南极洲以外均有蜘蛛分布，其种类多达50,000多种，大大小小，形态各异。

蜘蛛不属于昆虫。蜘蛛有八条腿，而昆虫有六条腿；蜘蛛的身体分为两部分，而昆虫的身体分为三部分；蜘蛛没有触角。不过，和昆虫一样，蜘蛛也属于节肢动物。蜘蛛有外骨骼（体表的铠甲），在幼蛛长大成为成蛛的过程中，必须定期蜕皮。（有时你会在蜘蛛网上看到蜘蛛蜕下的皮。）所有的蜘蛛都会纺丝，蛛丝从纺器中排出，很结实；纺器是位于腹部末端的小喷嘴状物。大多数蜘蛛有八只眼睛，也有些种类有六只、四只，甚至只有两只眼睛。

蜘蛛的身体包括两部分：头胸部和腹部。头胸部长有眼睛、口器和足；腹部有大多数内脏器官。

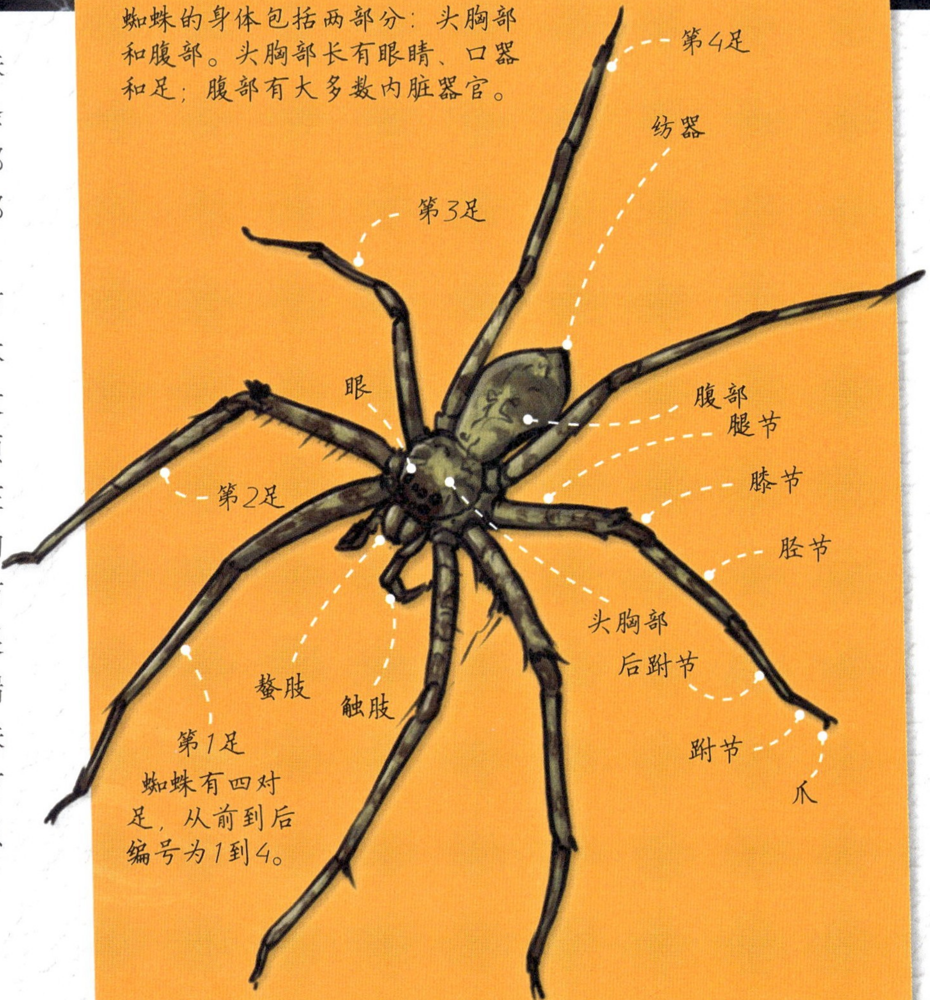

蜘蛛有四对足，从前到后编号为1到4。

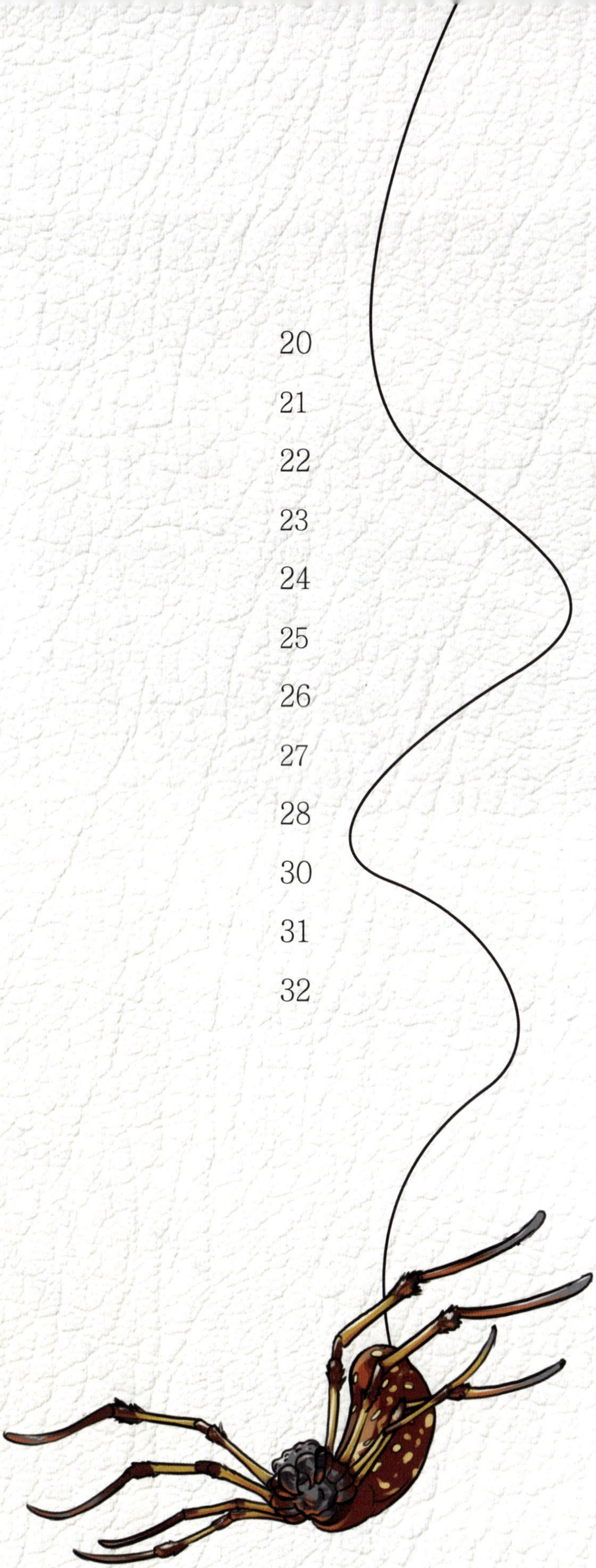

目 录

蜘蛛可以纺出强度甚于钢丝的蛛丝，这种绝技使得蜘蛛不仅可以织出优美的圆形蛛网，还可以建造精巧的陷阱、制作巧妙的绊线。这本书将带你近距离了解这些杰出的工程师、空中杂技演员和超级成功的捕食者。到这本书里，来探索：

- 能够拉动是自身重量 38 倍的重物的蝗蛸
- 一辈子生活在水里的潜水钟蜘蛛——当然它们也需要呼吸空气！
- 看起来像鸟粪或者甲虫的蜘蛛

……

当然还有更多令人着迷的蜘蛛奥秘等你探索。所以，下一次可要驻足看一看阳光下闪闪发光的蜘蛛网，还有那吊在蛛丝上的忙碌建筑师……

番茄博士双语百科

不可思议的虫子王国

超级蜘蛛

马特·特纳（英）著
圣地亚哥·卡列（英）绘
丛岚 译

外语教学与研究出版社
FOREIGN LANGUAGE TEACHING AND RESEARCH PRESS
北京 BEIJING

Record-Breaking Bugs

MAGNIFICENT MINI BUGS

Written by Matt Turner
Illustrated by Santiago Calle

外语教学与研究出版社
FOREIGN LANGUAGE TEACHING AND RESEARCH PRESS
北京 BEIJING

Some of the most powerful creatures are also the smallest! Some share our homes, such as the woodlouse and centipede; others are microbugs that can only be seen clearly under a microscope. Here you'll find

- fleas that can jump 30,000 times non-stop
- millipedes that make cyanide
- nits found on 3,000-year-old mummies

...

and many more tiny but tough customers. You will be brought face to face with these mini-monsters, such as the indestructible water bear, the sea spider which has guts in its legs, and ticks that taste with their toes. There's no escaping them!

CONTENTS

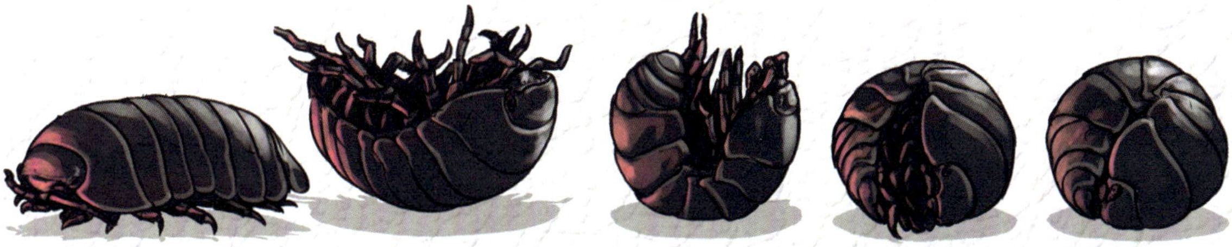

Magnificent Mini Bugs

Don't be fooled by size. Some of the most powerful creatures are also the smallest! This book looks at a few mini beasts. Some of these creatures you can see at home, such as the woodlouse and centipede; others are so small you need a microscope to study them. But all have powers of some sort: amazing defences, running fast, jumping high, surviving extreme environments…

Water bear

Flea

Unfortunately, when we look on the small scale, we find a lot of unpleasant beasts. Parasites, such as fleas and worms, live on or in the body of a 'host' animal, usually at the host's discomfort. Up to 50 per cent of all animal species are parasitic in some way. It's a good life for the parasite, after all, offering easy food and a high degree of protection.

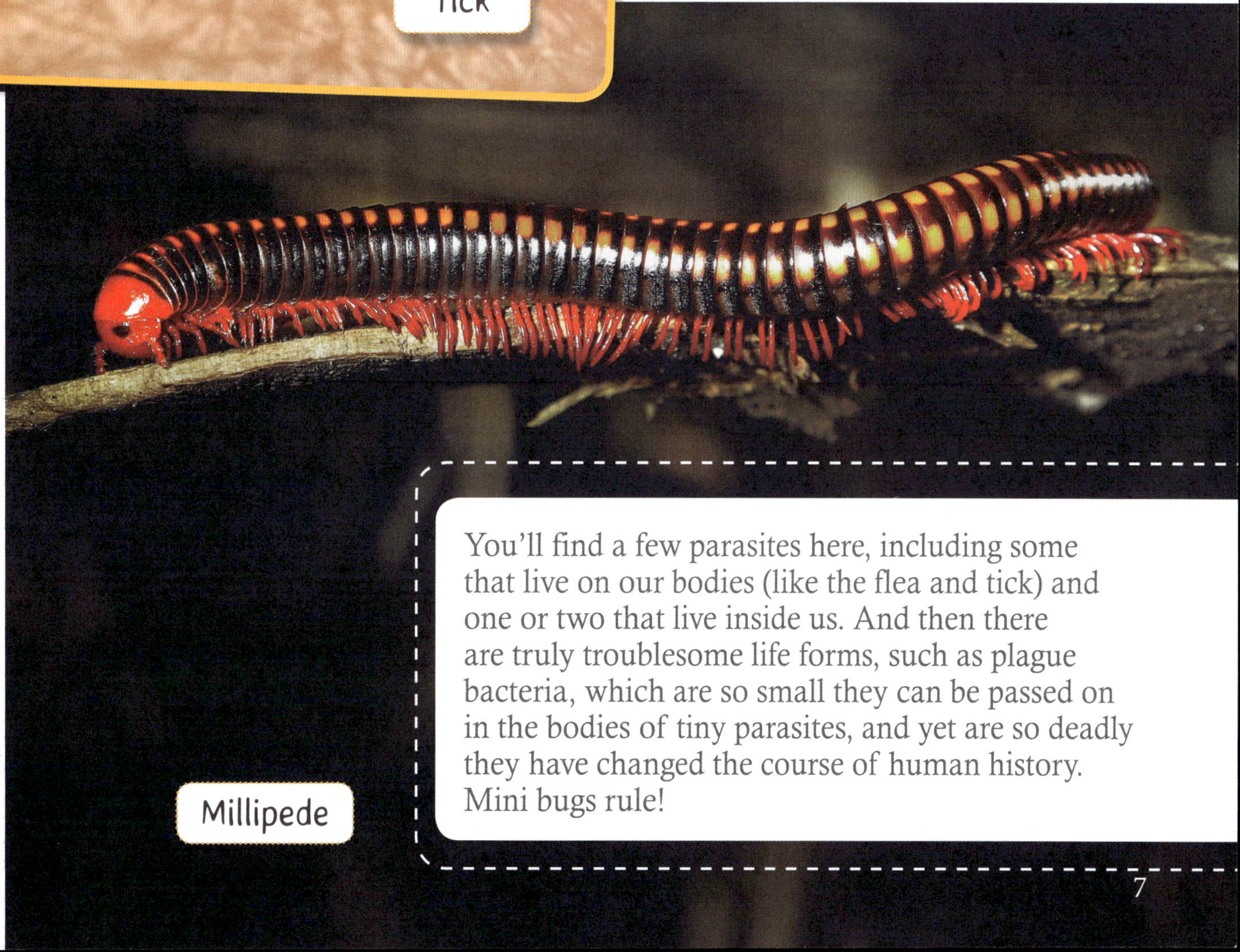

You'll find a few parasites here, including some that live on our bodies (like the flea and tick) and one or two that live inside us. And then there are truly troublesome life forms, such as plague bacteria, which are so small they can be passed on in the bodies of tiny parasites, and yet are so deadly they have changed the course of human history. Mini bugs rule!

Tough Critters

Water bears use body fluid pressure to bend their legs, which are tipped with sharp, curving claws.

When there's no water, water bears dry out and shrivel. In this state they are called 'tuns', and a sugar called trehalose replaces water in the body to preserve the cells.

Tuns usually survive drying out. They and their eggs may be carried on the wind to new places, to start new populations once they're moistened again.

In boiling and freezing tests, water bears have survived temperatures from –272 °C to 149 °C... way beyond human endurance!

In 2007, the European Space Agency sent tuns and eggs up on a rocket to expose them to the vacuum of space. They were also exposed to levels of solar radiation that would kill humans. After returning to Earth 10 days later and being rehydrated, two-thirds of the tuns survived the trip unharmed.

WATER BEARS

The tiny water bear is also called tardigrade, which means 'slow walker'. It clambers through damp plants, eating tinier animals or algae with its pump-action snout. And it's the toughest animal ever! In tests it has survived being boiled, frozen, pressurised, oxygen-starved and blitzed with radiation. But its best trick, used when there's no water, is to go into a tun state, when it shrivels like a dry sponge and simply waits – for years, sometimes – to be wet again.

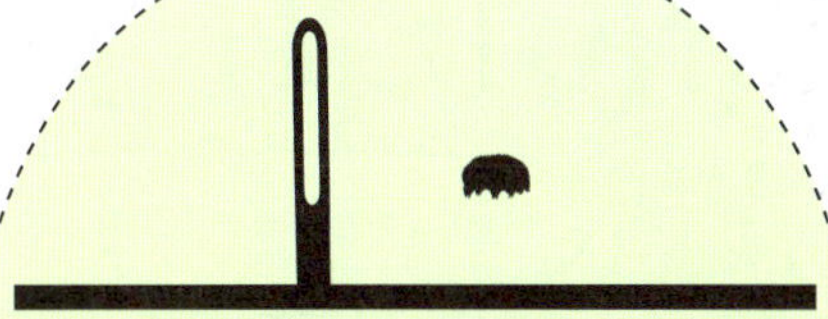

WATER BEAR

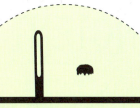

Lifespan: 10 years or more

Size: body length 0.1–1.5 mm

PUFFED UP

When starved of oxygen, water bears swell up (as shown above) until oxygen levels return to normal. This puffy state is called anoxybiosis.

ALIENS

People have suggested water bears might be aliens from another planet. But as they can go without oxygen for only a few days, this isn't likely.

SURVIVORS

Water bears can survive pressure almost six times greater than that found in the deepest ocean trenches.

Ancient Killers

The velvet worm group – the Onychophora – is very, very old. Fossil ancestors have been found that are half a billion years old.

Flanking the mouth of the velvet worm are two moveable turrets known as oral papillae. These face-guns squirt jets of slime up to 30 cm.

Velvet worms eat woodlice, spiders, crickets and more. After sneaking quietly up on prey, a velvet worm 'slimes' it to prevent escape. Then it uses the blade-like 'teeth' in its powerful jaws to chomp a hole in the victim's body. Finally, it injects the victim with saliva, which turns the guts to a goo it can suck up.

Velvet worms don't have a tough outer skeleton, but a soft covering called the cuticula. To grow, they shed this every couple of weeks or so.

Some species lay eggs. In others, a female gives birth – after a pregnancy of up to 15 months – to live young that can look after themselves from the start.

VELVET WORMS

Squashy, leggy and fuzzy to the touch, the velvet worm is not a worm, but a member of an ancient group of deadly predators. Living in damp leaf litter, it stalks a victim in total darkness, tapping it softly with its antennae to see if it's worth attacking. Then – splat! – the velvet worm shoots twin jets of slime from its face-guns, leaving the victim helpless to defend itself.

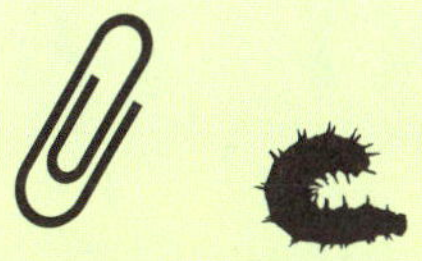

VELVET WORM

Lifespan: up to 6 years

Size: body length 15–150 mm

LOTS OF LEGS

Depending on species (about 180 worldwide), velvet worms have between 13 and 43 leg pairs, with claws that extend for extra grip on rough ground.

DARK DWELLERS

Velvet worms like darkness partly because they need to be damp, in order not to dry out. But if they get too wet, they drown.

HIERARCHY

Sometimes several velvet worms gather at a feed – but there's a strict pecking order as to who eats first, starting with the bossiest female.

Tiny Tanks

Whew, it's another scorcher today.

Whenever possible, woodlice squeeze their flattened bodies into tight spaces. One desert-dweller stays cool by digging a hole for itself, its partner and young.

Is there room for one more?

The sea slater, a cousin of woodlice, lives in the splash zone on the beach. It breathes air, hides in damp cracks and eats decaying seaweed.

Woodlice of the family Armadillidae are known as pill bugs or 'roly-polies'. They can curl their body up like a tiny armadillo when threatened by danger. You can find them in the garden. Their body is more rounded than a woodlouse's, but they're easily confused with pill millipedes, which are completely different animals.

Oooh, tickles!
You should try a hatchback.

Baby isopods, known as mancas, hatch in a pouch on the mother's underside. A pill bug (left) curls up to release her mancas; while a sea slater (right) raises her abdomen.

I hope they don't make ME try and balance a football on my nose.

A distant cousin of woodlice is the giant isopod *Bathynomus giganteus*, which can reach a whopping 36 cm! It is found in cold, deep waters.

WOODLICE

The 3,000-plus species of woodlouse are crustaceans: cousins of sea creatures like shrimps and crabs. Unlike any other crustacean group, they all live on dry land, in all sorts of habitats from deserts to mountains. But there's a chink in their armour: because their exoskeleton isn't waterproof, they can quickly dry out unless they find a dark, damp hiding-place – for example, under a plant pot in your backyard.

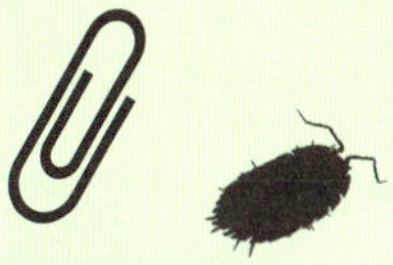

COMMON WOODLOUSE

Lifespan: usually 2 years, but may be up to 4
Size: body length about 16 mm

TWO COLOURS

Woodlice moult in two stages: first they unshell the back half, then a few days later, the front. That's why you sometimes see two-toned woodlice.

POO!

Woodlice eat their own poo, to recycle nutrients. And instead of peeing, they give off ammonia gas.

STALKING THE SEABED

Attached to the front end of a sea spider are a snout, palps and unusual claw-like organs called chelicerae. Not all species have the full kit, though.

The sea spider's abdomen is so thin, it has no room for guts, so these are located in the legs! It has no gills either, but takes in oxygen via its exoskeleton.

The long, thin legs are useful for sea spiders having to wade through seabed muck. Sea spiders in shallower zones tend to have stouter, stronger legs.

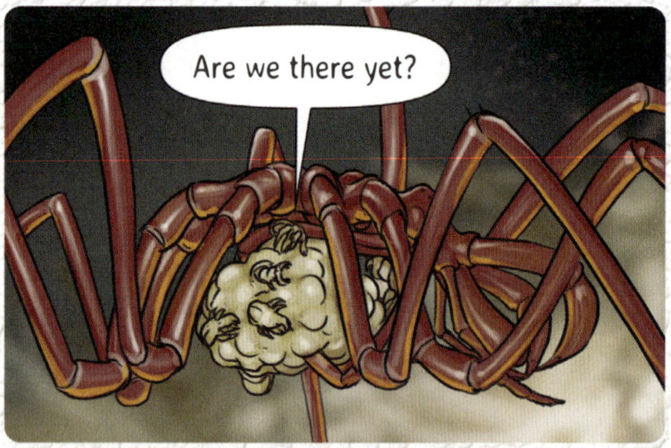

Most sea spiders have a pair of ovigers that are normally folded up against the body. The male sea spiders use ovigers to carry fertilised eggs.

Most sea spiders are smaller than a mosquito, but in the waters off Antarctica lurk giants measuring up to 60 cm across their legspan, and they rub shoulders with monster worms and crustaceans. It's thought that the high oxygen content of cold water helps support these larger-than-normal life forms.

SEA SPIDERS

With their eight legs, the sea spiders may look like land spiders, but they're an unrelated group of ancient marine arthropods. They are found worldwide, from shallow waters to deep ocean trenches. All sea spiders are scavengers or predators, with a long sharp snout for sucking a snack from the bodies of sea anemones and other seabed life. And, with a few exceptions, they're all unbelievably thin!

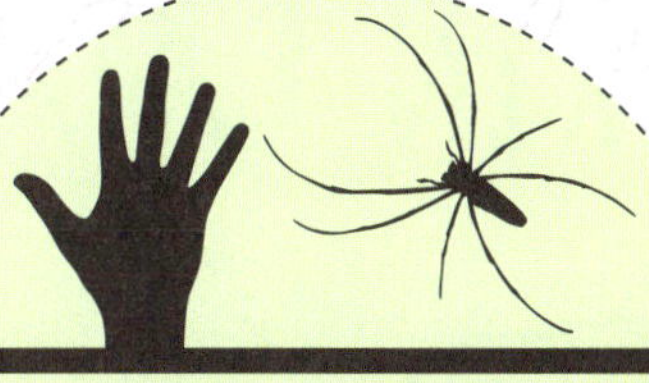

SEA SPIDER

Lifespan: not known

Size: legspan 1–700 mm

NO BREATHING

A typical sea spider is so skinny, it has no room for lungs or gills, so it doesn't breathe. Instead, the body absorbs oxygen directly from seawater.

TINY MUSCLES

The leg muscles of some sea spiders are microscopically small: they are formed from a single cell.

Skin Crawlers

There are two main types of louse. Some suck blood and other body fluids; others nibble on skin, feathers, dried blood and other surface grot.

One bird, the hooded pitohui of New Guinea, has toxic feathers and skin. Scientists think this may be an adaptation for keeping lice away.

Because nits (eggs) and adults cannot survive more than 24 hours away from a host's warmth, the lice use spit to glue their nits to hair or feathers.

There are flies that behave like lice. Louse flies have small (or no) wings, but cling tight to their host – such as a dog or bat – and suck its blood.

During World War I, thousands of soldiers suffered 'trench fever', a bacterial disease spread by lice that infested their clothing. Symptoms included a fever, headaches and leg pains. A favourite pastime was squishing the lice in their shirts.

LICE

Almost all mammals have something unpleasant in common. Lice! There are some 5,000 species of these tiny, wingless insects, and all are ectoparasites. They cling tight to the skin or hair of a host, and feed either by sucking its warm blood or chewing bits of dead skin. We humans are host to three species, which lay their nits (eggs) in our hair and on our clothes. Are you itching yet?

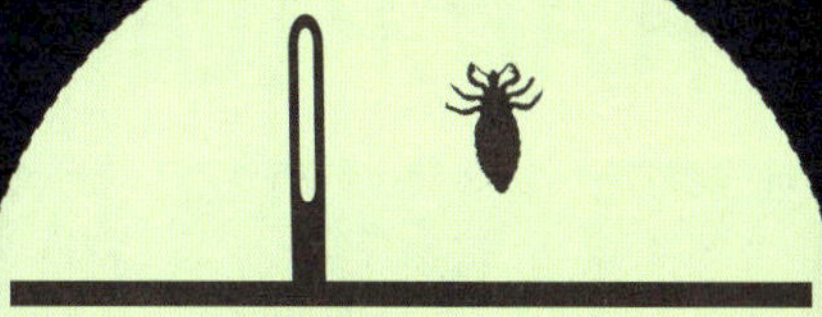

HUMAN HEAD LOUSE

Lifespan: 30 days from egg to death
Size: body length 2.5–3 mm

PARASITES

Bats and whales have no insect lice, but do have other parasites of their own. For example, whales have crustacean lice measuring up to 2.5 cm long.

NITS

Lice have probably always plagued humans. When archaeologists opened 3,000-year-old Egyptian tombs, they found nits on mummies.

CHAMPION JUMPERS

Ever seen a flea going backwards? Nope. Backward-pointing bristles on its body help it to push constantly – and rapidly – forwards like a rugby prop.

A flea can jump 30,000 times nonstop! It jumps by flexing a pad of resilin (a highly elastic protein), then releasing the energy through its legs.

Flea eggs, laid on the host, hatch into legless larvae, which eat the dried blood in the poo of adult fleas. Eggs and larvae may drop off a host at any time. The larva later pupates (develops inside a pupa), often resting in the carpet for months, till a host walks near. Then it suddenly springs out to jump on board.

Fleas breed so fast that, in just three weeks, a single adult pair can populate your pet with a thousand more fleas.

In 1330–1353, a plague known as the 'Black Death' killed more than 75 million people worldwide. The bacterium responsible was carried by fleas on rats.

FLEAS

These little wingless insects, tormentors of your pet cat or dog (or you), are perfectly designed. Powerful hind limbs launch them onto a host, and needle-like mouthparts stab the skin to release a blood meal. Like lice, fleas need their hosts. Adult fleas only live for a few days without feeding, so they snuggle in tight, and their slim, armour-plated bodies cannot be easily dislodged or crushed.

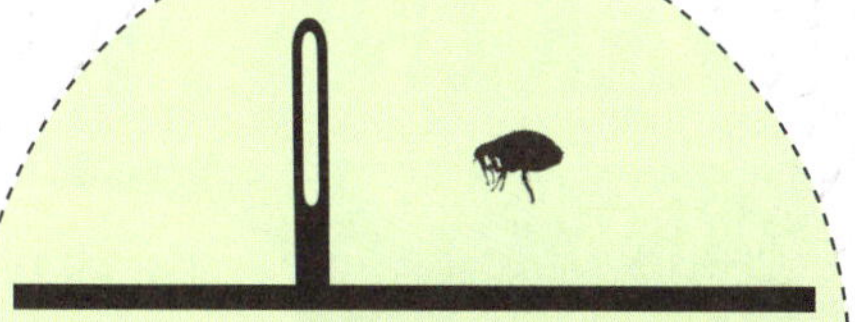

CAT FLEA

Lifespan: 1 month or more depending on conditions

Size: body length 1–2 mm

ANIMAL HOSTS

Rodents, rabbits, hoofed mammals, bats, birds… many different animals can be carriers of fleas.

SLIPPERY MOVES

Some fleas are self-lubricating. Pores in the body release an oily material that helps them slip quickly through fur.

Making You Itch

Mites infest animals of all sizes, from elephants to ants. These mites are using their ant host to 'hitch a ride' to somewhere new.

Dust mites love living in our pillows, where they can eat flakes of dead skin. Their poo can cause allergic reactions that make people sneeze and cough.

To get a blood meal from a host, such as a deer, a tick latches onto the body and bites into the skin. As it drinks, its abdomen swells up and darkens. After a female has fed, she drops to the ground and lays her eggs.

A tick can taste with its toes. It does this with an olfactory organ (known as Haller's organ) located in each of its front two feet.

Ticks spread serious diseases. Lyme disease, a bacterial infection that leaves you feeling stiff and tired (sometimes for years), is spread by the deer tick.

MITES AND TICKS

Both these creatures are eight-legged cousins of the spiders. Mites are found all over the world, and even in your bed – though you need a microscope to see them. Some eat organic matter (dead or living plants, dandruff, earwax…). Others are parasites on plants or on all kinds of animals, including humans. They burrow into the host's skin, feed on tissue and fluids, and cause misery. Ticks are easier to see – they are basically large mites.

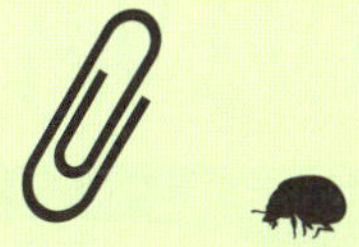

MITE

Lifespan: usually 1 month

Size: body length 0.1–6 mm

SKIN DISEASE

If you see a homeless dog scratching at rough, bare patches of skin, it's probably suffering from mange – an infection caused by mites.

HUMAN CONTACT

You can pick up ticks yourself from a walk in long grass. They must be removed carefully, so that the head isn't left buried in the skin.

SURVIVAL TRICKS

In defence, millipedes may coil up into a whorl, protecting the head. The masters at this are the pill millipedes, which look rather like pill bugs (page 12).

Many millipedes deter predators by oozing bad-tasting chemicals. Motyxia species glow, which is thought to be a warning that they contain cyanide.

Meerkats and coatis are mammals that eat millipedes; they roll them on the ground to 'detox' them first. And some monkeys rub the noxious secretions into their coat as a knockout mosquito repellent!

Some soft-bodied millipedes have tufts of barbed bristles, which they wipe off against enemies, such as ants, to tangle them up.

Millipedes might cause train crashes by swarming over rail tracks. When the wheels squish them, they tend to slip and derail.

MILLIPEDES

These harmless vegetarians feed on decaying matter on the forest floor, ooching along by means of waves that ripple along their many legs. Harmless? Well, not quite. To fend off predators, they can ooze – or, in some cases, squirt – such nasty toxins as hydrogen cyanide, hydrochloric acid and benzoquinone, which can burn your skin and dye it brown. Handle with care!

MILLIPEDE

Lifespan: up to 10 years or more
Size: body length 2–380 mm

LEG COUNT

Millipedes are diplopods, with two pairs of legs per body segment. Depending on species, the leg count varies from 24 all the way up to 750.

FOSSIL PROOF

Fossils with ozopores – the tiny holes from which millipedes leak their toxins – prove these chemical defences are at least 420 million years old.

Leggy Hunters

Argh! To the bat cave!

Centipedes like to be in a tight spot: they are happiest when both their upper and lower surfaces are touching something firm.

Ah, food!
Oh no, wait – that's just me.

Centipedes don't mind a bit of cannibalism and will readily eat other centipedes, especially if they find an injured one that cannot escape.

I'll work my way up to the bat – gradually.

Some centipedes can raise their forequarters into the air and catch bees or wasps in flight! The Peruvian giant centipede is a 30-cm-long monster that has perfected the art of hanging from cave roofs. Despite having poor eyesight, it can catch bats in flight as they leave their roost in the evening.

Hey, stop wriggling – that tickles!

Many centipedes are model parents. For example, a mother may lick her eggs to keep them free of fungi, or curl her body protectively around hatchlings.

Ha! Tweet that!

If caught by the legs, a centipede can 'drop' them. While the wriggling legs distract the attacker, the animal escapes. New legs grow after moulting.

CENTIPEDES

Running on up to 360 legs, centipedes are super-fast, fleeing from view if you uncover them in the yard. Sunlight dries them, so they usually spend the day in a moist hiding-place, coming out at night to hunt. Armed with venom-packed pincers, centipedes prey on anything they can catch. Small species hunt flies and beetles; tropical giants may tackle birds, lizards and mice – and give humans a nasty bite, too.

CENTIPEDE
Lifespan: up to 10 years
Size: body length 10–300 mm

HYGIENE
After feeding, centipedes carefully clean their antennae and legs – all of them! – by running them through the mouthparts.

IN A DASH
The American house centipede can cover 40 cm in a second. That's more than 10 body lengths.

Incredible Journey

The liver fluke's life cycle begins when an adult, living in a mammal's liver, sheds eggs. These come out in the mammal's droppings, such as a cowpat.

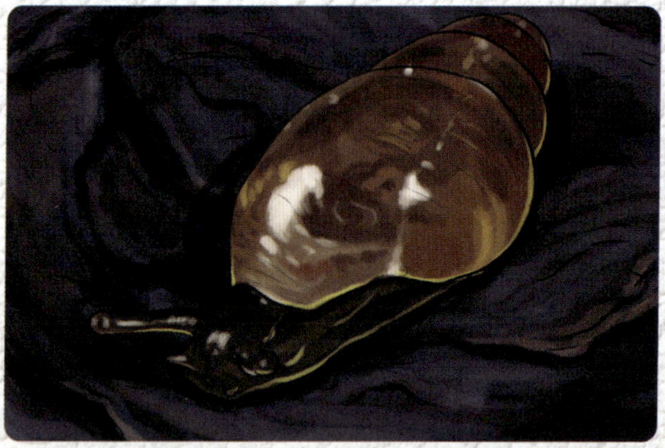

A snail eats some cow poo and, with it, some fluke eggs. The eggs then develop into tiny forms called cercariae, which multiply inside the snail.

Irritated by the cercariae, the snail coughs them up in a slime ball, which is later discovered by an ant. The thirsty ant consumes some slime, and with it, some cercariae. These cercariae travel into the ant's brain and control its behaviour, so that it becomes a mindless zombie.

Every evening, the ant climbs a blade of grass and grips the tip with its jaws, then just sits there. Hours later, it returns to its colony, but repeats this night after night.

Eventually, a cow chomps the grass stem and swallows the ant with its parasitic cargo. The adult liver flukes develop and burrow into the cow's liver, in time laying eggs of their own.

FLUKES

Flukes are flatworms, belonging to a class called Trematoda. They are parasites that enter the guts of snails, fish and birds, as well as sheep or cattle, where they feast on the host's body fluids. Flukes can make livestock so sickly that farmers sometimes lose entire herds. How the flatworms enter those bodies is truly astonishing as it involves not one, but several hosts…

LANCET LIVER FLUKE

Lifespan: dependent on host
Size: adult body length up to 15 mm

INFESTATIONS

Humans can suffer from fluke infestations, too. One way to 'catch' them is by eating unwashed watercress or undercooked meat.

KILLERS

One group of trematodes, the blood flukes, is responsible for killing up to 200,000 people each year, mostly in Africa, Asia and South America.

Mini Whatsits

Mini bugs, though small, have standout qualities — whether it's their size, their lifestyle, or simply how long they've been around on planet Earth, as these examples show.

Mini-pede:
The smallest millipede is probably *Polyxenus lagurus*, found in North America and Europe. Measuring just 3 mm long, it is of the type that can shed bristles in self-defence. The longest is probably *Archispirostreptus gigas* of East Africa (left), which can reach 380 mm. The record for most legs – of any animal – goes to the tiny *Illacme plenipes* of California, which has up to 750 legs but measures barely 25 mm long.

Mite be useful: The mites on this sexton beetle are hitching a ride — and the host puts up with them because they're useful. The beetle lays its eggs in animal carcasses, such as dead mice, for its babies to feed on. The mites can usually be relied on to eat the eggs and larvae of any other species (such as a fly) that has reached the carcass first.

Super slimy:

Planarians are a group of predatory flatworms; some species live in water, others on land, where they move like slugs on a layer of slime. If a planarian is cut into two or more pieces, each piece grows back all of its missing parts to become a complete individual. Even if you chop the head off, it'll grow a whole new body. How's that for a super power!

Micro-shrimp:

These fingernail-sized crustaceans are brine shrimps. Their eggs, which can wait for up to 25 years before hatching into healthy larvae, are cultivated widely for use as farmed fish food. Like water bears, brine shrimps have been sent experimentally into space, but with a much lower survival rate.

Ancient insect:

This ancient dragonfly is trapped in amber. Its ancestors first appeared around 400 million years ago. They first evolved winged flight about 300 million years ago, which is when land plants began to grow much taller. Scientists reckon insects and plants evolved together. Today, around 65 per cent of plants are insect-pollinated.

Parasites on Parade

Many mini beasts are parasites: creatures that depend on other creatures for their survival – perhaps to gain free food or shelter. Below are a couple more parasites you may find interesting…

Leeches

Leeches are segmented worms with a sucker at each end of a muscular, stretchy body. With around 680 species known worldwide, they range from 7 to 300 mm. Most live in sluggish fresh water, where they latch onto almost any creature – living or dead – to suck its fluids. They often go for snails, but if you're unlucky a leech may find you. If you don't remove it first, it'll drop off after drinking its fill of blood.

When biting, leeches inject a chemical – hirudin – that prevents blood clotting, so wounds can 'leak' for hours after the leech has been removed. For more than 2,000 years, drawing blood was a common cure for a wide range of ailments, and doctors put leeches on their patients to do the blood-letting. Some still do!

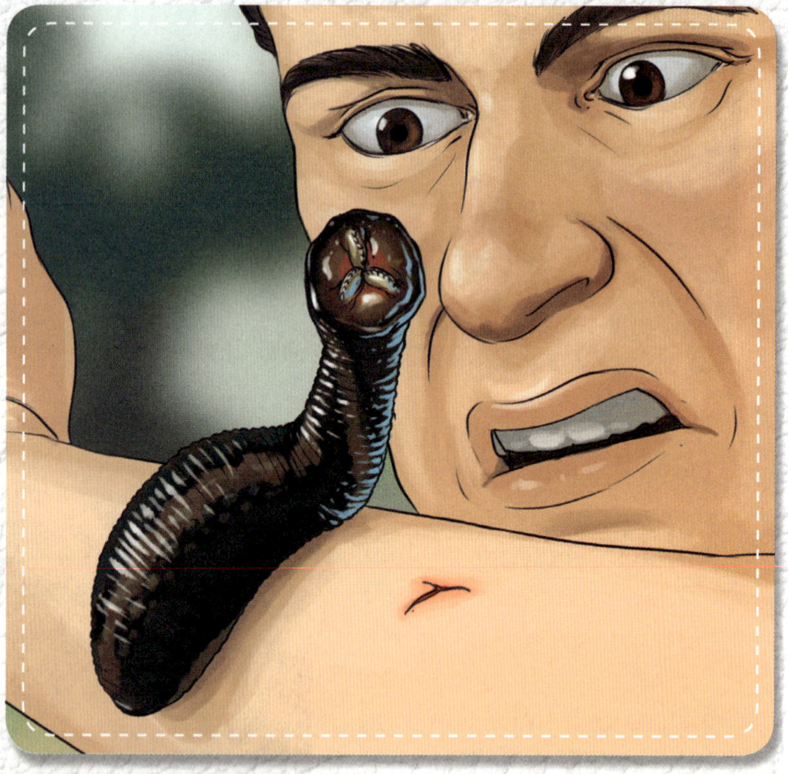

Tapeworms

Tapeworms belong to a phylum called Platyhelminthes (meaning 'flatworms'). One type commonly found in humans is *Diphyllobothrium*, which you can pick up by eating raw or undercooked fish infested with the tapeworm larvae. The tapeworm then grows in your gut, where it feeds by absorbing nutrients; in other words, it eats your food. Tapeworms don't really qualify as 'mini' beasts, given that they can rapidly grow to terrifying lengths. The longest ever found in a human gut was 25 m long! But don't worry: you can usually get rid of them with tablets.

Glossary

abdomen the hind part of an insect's three-part body (along with the head and thorax).

arthropod a member of the very large invertebrate group that includes crustaceans, insects, myriapods (centipedes, millipedes, and so on) and spiders.

bacterium (plural: bacteria) a microscopically small life form, often occurring in soil or water (or bodies) in large numbers, and some are capable of causing sickness or disease.

class a principal taxonomic category that ranks above order and below phylum, such as Mammalia (mammals).

crustacean a member of the arthropod group that includes barnacles, crabs, lobsters and shrimps. Most are marine, but some, such as woodlice, live on land.

exoskeleton the 'shell' or outer casing that provides structural strength and protection for an arthropod, in place of an internal skeleton. Usually made of chitin, the exoskeleton must be moulted regularly for an arthropod to grow.

larva (plural: larvae) juvenile that hatches from the egg and later transforms into a pupa, or directly into an adult.

moult in insects, to shed the old exoskeleton (outer part) in order to grow.

olfactory the sensory system used for the sense of smell.

papilla (plural: papillae) a tiny, rounded bobble on an organ, often having a sensory function. Your tongue, for example, is covered with papillae – these are the taste buds.

parasite a life form that spends all or part of its life cycle on or inside another life form, which is known as the host. An obligate parasite cannot survive away from its host. An ectoparasite lives on the outside (skin, fur, and so on); an endoparasite lives on the inside, often in the digestive system.

phylum a principal taxonomic category that ranks above class and below kingdom. A phylum contains different classes. Mammalia belongs to a phylum called Chordata.

predator an animal that kills and eats other animals. Roughly one-third of all insect species are predatory.

pupate to transform into a pupa, a stage of metamorphosis in which the insect larva rests inside a case and gradually transforms into an adult.

species the basic unit of biological classification. Members of a species are defined as a group of individuals that are similar enough to be able to breed and produce fertile offspring.

vacuum a space that contains no air, and therefore no air pressure.

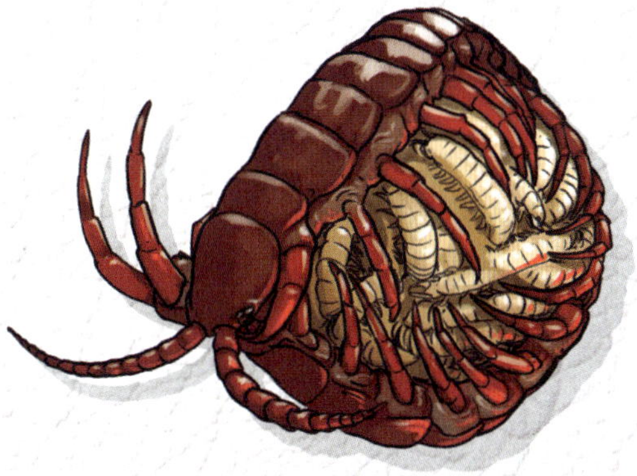

INDEX

The Author

British-born Matt Turner graduated from Loughborough College of Art in the 1980s, since which he has worked as a picture researcher, editor and writer. He has authored books on diverse topics including natural history, earth sciences and railways, as well as hundreds of articles for encyclopedias and partworks, covering everything from elephants to abstract art. He and his family currently live near Auckland, New Zealand, where he volunteers for the local Coastguard unit and dabbles in art and craft.

The Artist

Born in Medellín, Colombia, Santiago Calle is an illustrator and animator trained at Edinburgh College of Art in the UK. He began his career as a teacher, which led him to deepen his studies in sequential art. Santiago founded his art studio Liberum Donum in Bogotá in 2006, partnering with his brother Juan. Since then, they have dedicated themselves to producing concept art, illustration, comic strip art and animation.

Picture Credits (abbreviations: t = top; b = bottom; c = centre; l = left; r = right)
© www.shutterstock.com:

1 c, 2 cl, 4 c, 6 tr, 6 bl, 7 tl, 7 tr, 7 b, 9 c, 11 c, 13 c, 15 c, 19 c, 21 c, 23 c, 25 c, 27 c, 28 tl, 28 br, 29 tl, 29 cr, 29 bl, 32 cr.

17 c © CDC/Frank Collins, Ph.D/James Gathany

索引

作者简介

马特·特纳出生于英国，20世纪80年代毕业于拉夫伯勒大学艺术学院，毕业后一直担任图片研究员、编辑和作者。他的书题材广泛，涉及博物学、地球科学和铁路等，并为百科全书和分册出版的丛书写过数百篇文章，从大象到抽象艺术无所不包。他现在和家人住在新西兰奥克兰附近，他还是当地海岸警卫队的志愿者，平时也涉猎工艺品的制作。

绘者简介

圣地亚哥·卡列出生于哥伦比亚的麦德林，是一位插画师和动画师，曾在英国爱丁堡艺术学院接受过培训。他的第一份职业是教师，教学促使他在连续性艺术领域继续深造。2006年，他和兄弟胡安在波哥大合伙创立了一家名为“自由德南”的艺术工作室，自此两人便致力于概念艺术、插画、连环画和动画的创作。

Picture Credits (abbreviations: t = top; b = bottom; c = centre; l = left; r = right)
© www.shutterstock.com:

1 c, 2 cl, 4 c, 6 tr, 6 bl, 7 tl, 7 tr, 7 b, 9 c, 11 c, 13 c, 15 c, 19 c, 21 c, 23 c, 25 c, 27 c, 28 tl, 28 br, 29 tl, 29 cr, 29 bl, 32 cr.

17 c © CDC/Frank Collins, Ph.D/James Gathany

术语表

捕食者　捕食其他动物的动物。大约三分之一的昆虫都是捕食者。

腹部　昆虫身体分为头、胸、腹三部分，腹部是最后面的一部分。

纲　生物分类系统中的一个阶元，介于“门”和“目”之间，比如哺乳纲（包括所有哺乳动物）。

化蛹　变态发育的昆虫从幼虫形成蛹的阶段。昆虫的化蛹期在蛹壳里度过，在此期间，成虫结构逐渐形成。

寄生虫　在全部或部分生命周期内，寄生于另一个生物体内或体表的生物，被寄生的生物为其宿主。其中，专性寄生虫离开宿主无法生存。体外寄生虫寄生在宿主的体外，比如皮肤、毛皮等；体内寄生虫寄生在宿主的体内，通常是在消化系统里。

甲壳动物　节肢动物的一个类群，包括藤壶、螃蟹、龙虾和小虾在内。甲壳动物多生活在海洋里，也有一些在陆地上生活，比如鼠妇。

节肢动物　无脊椎动物的一大类群，包括甲壳动物、昆虫、多足纲节肢动物（蜈蚣、马陆等）和蜘蛛在内。

门　生物分类系统中的一个阶元，介于“界”和“纲”之间。同一门下有不同的纲。哺乳纲属于脊索动物门。

乳突　分布在器官表面的细小的圆形小球状突起，通常具有感觉功能。例如，舌头表面就密集分布着许多的小球状突起，这就是味蕾。

蜕皮　对于昆虫而言，蜕皮就是脱去旧的外骨骼，以适应生长的需求。

外骨骼　节肢动物体表的外壳。它代替内骨骼，起到结构支撑和保护作用，通常为甲壳质。节肢动物在生长过程中必须定期蜕去外骨骼。

细菌　一种极其微小的生命形式，需借助显微镜观察，通常大量存在于土壤、水和生物体内。有些细菌能引发疾病或不适。

嗅觉器官　感受嗅觉的感官系统。

幼虫　从卵孵化出来的幼体，在幼虫期结束后会化蛹或直接发育成成虫。

真空　没有空气也没有大气压力的空间。

种　生物分类的基本单位。同一种的个体之间具有足够相似的特征，可以相互交配并产出有繁殖能力的后代。

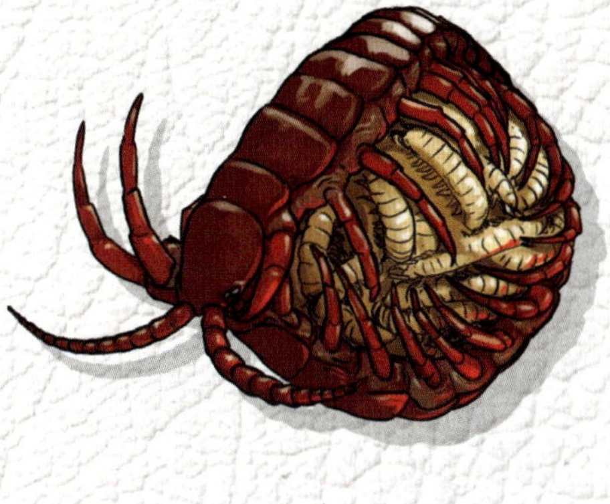

寄生虫展览馆

许多小虫子都是寄生虫，也就是说，它们依赖其他生物生存，或是为了获取免费食物，或是为了获得庇护。下面再介绍一些你可能会感兴趣的寄生虫……

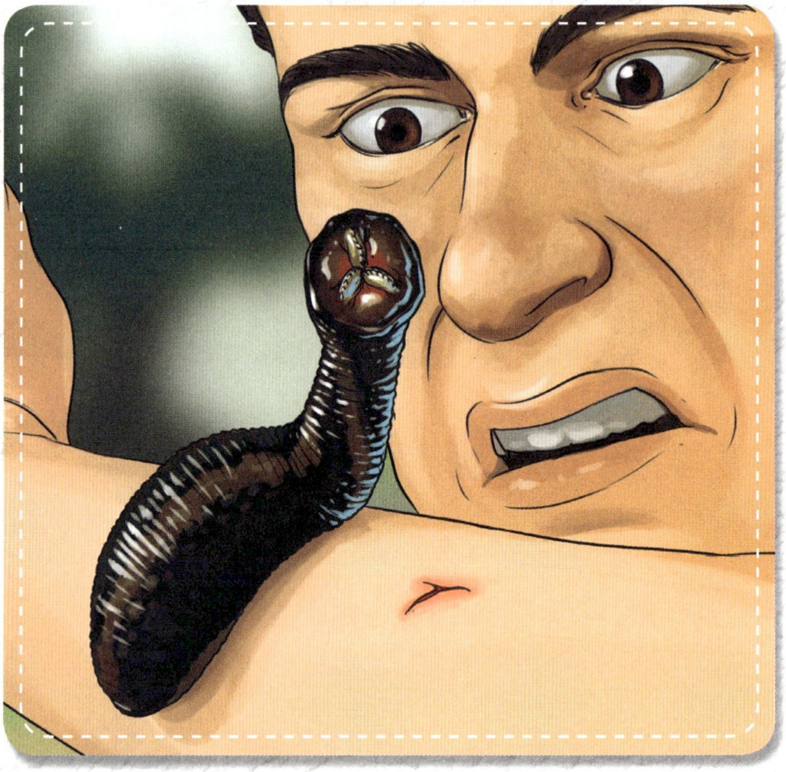

水蛭

水蛭是环节动物，躯体柔韧可伸缩，两端各有一个吸盘。全世界范围内已知的水蛭种类大约有680种，体长从7毫米到300毫米不等。大多数水蛭生活在流速缓慢的淡水中，吸食几乎所有动物的体液，无论对方是死是活，都能被它缠上。水蛭经常袭击蜗牛，但如果你不走运，水蛭也可能会盯上你。如果你没在第一时间把它弄走，它吸饱血后才会自动脱落。

水蛭在叮咬时，会注入一种名叫“水蛭素”的化学物质以防血液凝固。所以在清除水蛭数小时后，伤口仍会渗血。2,000多年来，人们治疗很多疾病都会采用放血疗法，医生曾把水蛭放在病人身上进行放血。现在也有人使用这种方法！

绦虫

绦虫属于扁形动物门，是扁虫的一种。常见于人体中的是裂头绦虫，如果你吃了生的或者半生不熟的鱼肉，里面又碰巧寄生了裂头绦虫的幼虫，就可能会染病。接下来，裂头绦虫会寄生在你的肠道里，一点点长大，吸收肠道中的营养，换句话说，就是吃掉你肚子里的食物。绦虫真的不应该算是小虫子，因为生长迅速，用不了多久，它们的体长就能吓你一大跳。有史以来在人类肠道中发现的最长的绦虫竟然长达25米！但是别担心，通常来说，你吃点药就能将绦虫排出体外。

超级黏滑

涡虫是一种扁虫，还是一种捕食性动物。其中有些种类生活在水中，还有些生活在陆地上。陆地上的涡虫像蛞蝓一样，在黏液上滑行。涡虫被切成两段或多段后，每一段都能再生出缺失的部分，成为一只完整的个体。即使是脑袋被砍掉了，它也能重新长成一只全新的涡虫。这项超能力很酷吧！

显微镜下才能看到的虾

这些指甲盖大小的甲壳动物是丰年虾，它们的卵可以保存25年之久，一旦条件具备，便能孵化成健康的幼虫，所以被广泛养殖，用做鱼类的饵料。像水熊虫一样，丰年虾也曾被送到太空进行实验，但是它们的存活率相比水熊虫要低得多。

古老的昆虫

这只古老的蜻蜓被困在琥珀中。蜻蜓的祖先最早出现在大约四亿年前。大约三亿年前，当陆地上的植物开始变高时，它们进化出翅膀，学会了飞行。科学家认为，昆虫和植物是同步进化的。今天，大约有65%的植物是虫媒传粉植物。

小虫子档案室

小虫子个头虽小，却具有令人惊奇的特质，无论是它们的大小、生活方式，还是在地球上的生存时间，都让人惊叹。请看下面的例子。

马陆之最

体形最小的马陆可能是兔草毛马陆。它们分布在北美洲和欧洲，体长只有3毫米，能自主脱落刚毛用于自卫。身体最长的马陆可能是来自东非的非洲巨马陆（见左图），体长可达380毫米。腿最多的马陆当属分布在加利福尼亚州的全真千足虫，它们身体很小，只有25毫米长，却有多达750条腿，这在所有的动物当中都是绝无仅有的。

螨虫好帮手

这只埋葬虫让身上的螨虫搭了个便车。埋葬虫作为宿主，之所以能够容忍螨虫，是因为螨虫是它的好帮手。埋葬虫把卵产在老鼠等动物的尸体里，使这些尸体供孵化后的幼虫食用。但苍蝇等其他物种可能会捷足先登，埋葬虫便常常依赖螨虫吃掉这些动物的卵和幼虫。

吸虫

吸虫是一种扁虫，属于吸虫纲。吸虫还是一种寄生虫，它们进入蜗牛、鱼类、鸟类、绵羊或牛的内脏，尽情享用宿主的体液。吸虫会使牲畜患上严重的疾病，甚至成群死亡。这些扁虫进入宿主体内的过程让人瞠目结舌，因为其中涉及的宿主不是一个，而是多个……

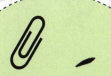

矛形双腔吸虫

寿命：取决于宿主

大小：成虫体长可达15毫米

侵扰人类

人类也会遭受吸虫侵袭。一种感染途径是吃未洗过的豆瓣菜或未做熟的肉。

血腥杀手

因血吸虫（吸虫的一种）而导致的死亡人数每年多达20万，发病地区主要是非洲、亚洲和南美洲。

神奇的生命旅程

肝吸虫的生命周期始于成虫在哺乳动物的肝脏内产卵。随后，虫卵会随着哺乳动物的粪便（如牛粪）排到体外。

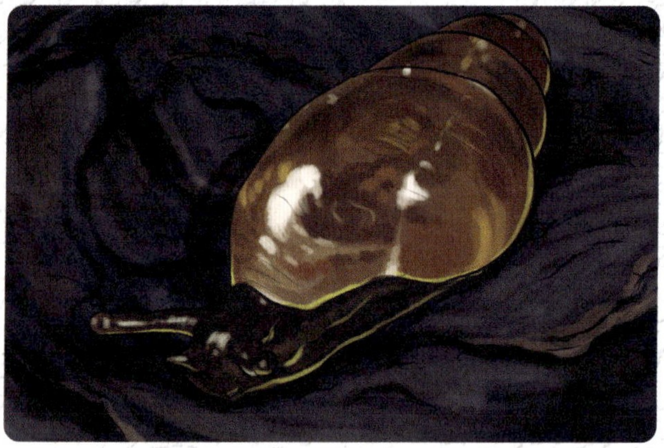

蜗牛吃下牛粪的同时，也将虫卵吃进了肚子。接下来，虫卵发育成微小的尾蚴，并在蜗牛体内大量繁殖。

尾蚴的刺激引起蜗牛咳嗽，蜗牛将尾蚴裹在黏液球里咳出来。之后，一只口渴的蚂蚁发现了黏液球，它吸食黏液解渴时，尾蚴便乘机而入。接下来，这些尾蚴进入蚂蚁的大脑，控制蚂蚁的行为，使它成了一具无意识的僵尸。

每天傍晚，这只蚂蚁都会爬到一片小草叶子上，用颚部紧紧夹住叶尖，待在叶片上发愣。几小时后，它会回到蚁群。每个夜晚，它都会重复这一行为。

最终，一头牛来吃草，吞下寄生了肝吸虫的蚂蚁。肝吸虫进一步发育，钻进牛的肝脏，并适时产下虫卵。

蜈蚣

蜈蚣的腿多达360条，所以跑起来超级快。你在院子里发现它们的一瞬间，它们就会从你的视线中消失。为了避免被晒成“蜈蚣干”，它们通常白天躲在潮湿的隐蔽处，晚上再出来觅食。凭着那对带有毒液的“钳子”，蜈蚣所向披靡，抓到什么吃什么。小型蜈蚣捕食苍蝇和甲虫，热带的巨型蜈蚣则会攻击鸟类、蜥蜴和老鼠，甚至还会狠狠咬人一口。

蜈蚣

寿命：长达10年

大小：体长10—300毫米

有洁癖

进食后，蜈蚣会仔细地用口器清洁自己的触角和腿——一个都不落下！

速度之王

美洲蚰蜒（蜈蚣的一种，译注）1秒可以爬行40厘米，相当于身长的十倍多。

我的腿数不清

蜈蚣喜欢待在狭窄的空间里，最喜欢将腹背两侧都紧贴着坚固的物体。

蜈蚣毫不介意同类相食，甚至还挺乐意，尤其是当同类受了伤、无法逃脱时，它们会吃得更爽快。

有些蜈蚣能把前半截身体抬到空中，捕捉飞行的蜜蜂和胡蜂！秘鲁巨型蜈蚣身长30厘米，这种怪物虽然视力不好，却能完美地倒挂在洞穴顶部，在蝙蝠夜晚从窝里飞出来的时候将它们擒获。

许多蜈蚣是模范父母。例如，一只蜈蚣妈妈可能会舔舐自己的卵，使其免受真菌感染，或者蜷着身体，将刚孵化出的蜈蚣宝宝们裹在身下来保护它们。

当蜈蚣的腿被逮住时，它们可以自断腿部。扭动的蜈蚣腿可以分散攻击者的注意力，让蜈蚣趁机逃脱。蜕皮之后，它们会长出新腿。

马陆

马陆是无害的素食者，以林地表层的腐烂物质为食，前后足依次前进，形成波浪式运动。无害？嗯，也不尽然。为了防御捕食者，它们会分泌或者喷射气味难闻的毒素，比如氰化氢、盐酸和苯醌，这些毒素能灼伤你的皮肤，将皮肤染成褐色。小心轻触！

马陆

寿命：长达10年及以上

大小：体长2—380毫米

马陆有多少条腿？

马陆是倍足纲动物，每个体节上有两对足。不同种类的马陆，足的数量有别，从24条到750条不等。

化石证据

马陆通过一些叫做“臭腺孔”的小孔分泌毒素。带有臭腺孔的化石证明，这种化学防御方式至少已有4.2亿年的历史了。

生存大师

马陆在御敌时会把身子蜷成螺旋形，以保护头部。球马陆最擅长这一招数，对，就是那些看起来酷似球潮虫的家伙（参见第12页）。

许多马陆通过分泌味道糟糕的化学物质来震慑捕食者。加州山蛩虫（马陆的一种，译注）会发光，那似乎是在警告捕食者："我体内有氰化物，别惹我！"

有些哺乳动物吃马陆，比如獴和中南美洲浣熊。在吃之前，它们会将马陆放在地上滚一滚，排出毒素。有些猴子还把马陆的毒性分泌物蹭在身体表面，作为效果奇强的驱蚊剂！

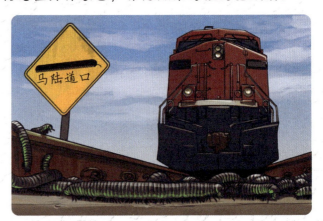

有些身体柔软的马陆长着一簇簇带倒钩的刚毛。在防御蚂蚁等天敌时，它们会用刚毛把对方缠起来。

马陆在成群通过铁轨时，可能会造成火车事故。当火车碾过马陆的时候，车轮可能会打滑，容易引起脱轨。

螨虫和蜱虫

螨虫和蜱虫都有八条腿，和蜘蛛是近亲。螨虫遍布世界各地，甚至还会光顾你的床——当然，你需要用显微镜才能观察到。它们当中有的吃有机物，包括死的或活的植物、头皮屑、耳屎……有的寄生在植物上或各种动物（包括人类）身上。它们钻入宿主的皮肤，以机体组织和体液为食，让宿主痛苦不堪。其实，蜱虫基本上可以说是大型螨虫，所以更容易被发现。

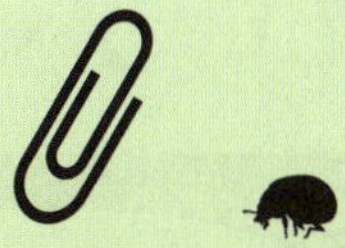

螨虫

寿命：通常为1个月

大小：体长0.1—6毫米

引发皮肤病

如果你看到一只流浪狗抓挠皮肤上粗糙的秃块，那它很可能是患上了兽疥癣，这是由螨虫感染引起的一种皮肤病。

与人类接触

当你在茂密的草丛中散步时，有可能会被蜱虫咬到。一旦被蜱虫叮咬，务必小心翼翼地将它们移除，以免将蜱虫的头部留在皮肤里。

挠痒痒高手

螨虫的动物宿主大小各异，大到大象，小到蚂蚁。上图中这群螨虫正在搭蚂蚁宿主的便车去往新的地方。

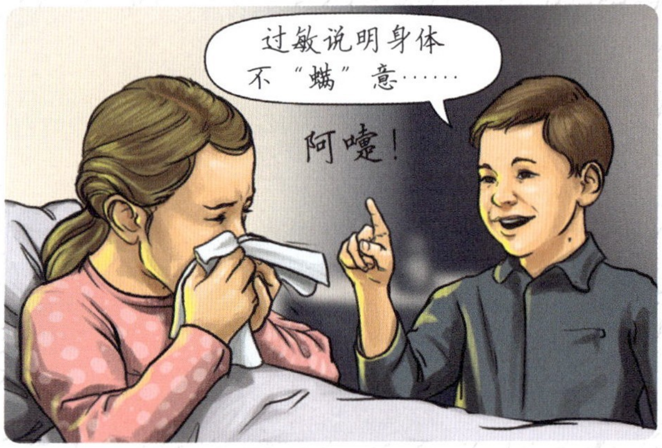

尘螨喜欢待在枕头里吃些死皮碎屑。它们的粪便可引起打喷嚏、咳嗽等过敏反应。

蜱虫想从鹿这样的宿主身上获取血液大餐时，需要紧紧地附着在宿主身体上，咬穿其皮肤。蜱虫吸血时，腹部会膨胀，颜色会变暗。雌蜱吸饱血后会落到地上产卵。

蜱虫能用足尖感受气味，它的第一对前足上长有被称为“哈氏器”的嗅觉器官。

蜱虫可以传播重病。莱姆病是以鹿蜱为媒介的一种细菌感染性疾病，患者感到身体僵硬、疲倦，症状有时会持续好几年。

跳蚤

这些没有翅膀的小昆虫让你的爱猫、爱狗甚至是你本人吃尽苦头。跳蚤的身体拥有完美的设计：强健有力的后腿让它能够弹跳到宿主身上；针状的口器帮助它刺穿宿主的皮肤，吸食血液大餐。跳蚤和虱子一样，需要依附宿主生存。成虫在不进食的条件下只能存活几天，所以它们紧紧地附着在宿主的身体上。跳蚤身体细小，还套着一层硬壳。你想把它们撵走或者压扁？没那么容易！

猫蚤

寿命：1个月及以上，
取决于周围条件
大小：体长1—2毫米

动物宿主

啮齿动物、兔子、有蹄哺乳动物、蝙蝠、鸟类等，都有可能成为跳蚤的宿主。

走路滑溜溜

有些跳蚤有自我润滑的本事。它们可以通过身上的毛孔分泌一种油性物质，从而敏捷利落地穿行于宿主的皮毛之间。

跳高冠军

我不会倒着跳。

你见过跳蚤倒着跳吗？谁都见不到。跳蚤身上的刚毛向后生长，可以推动它不停地向前快速运动，就像英式橄榄球中的支柱前锋一样。

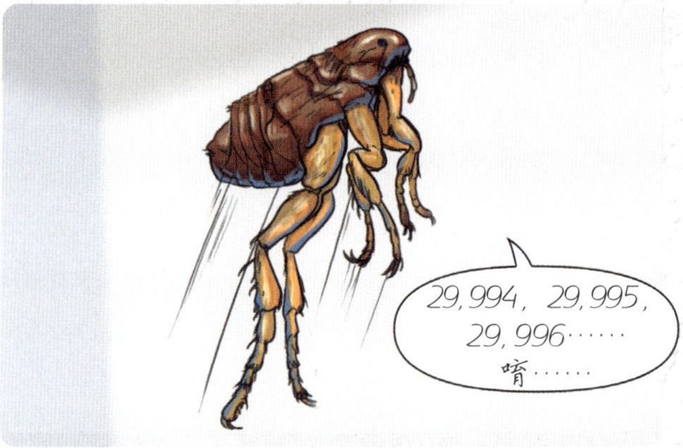
29,994，29,995，
29,996……
唷……

跳蚤能连续跳跃30,000次！它收缩体内的节肢弹性蛋白（一种高弹性的蛋白质）块，然后通过腿部释放能量完成跳跃。

早餐来啦！

跳蚤将卵产在宿主身体上，孵化出来的幼虫没有腿，以成虫粪便中的干血为食。卵和幼虫随时都可能从宿主身上掉下去。一段时间过后，幼虫开始化蛹，在蛹里进一步发育成成虫。成虫守株待兔，往往在地毯上一等就是好几个月，直到宿主走近，刹那间它会腾空而起，跳到宿主的身上。

孩子们都去
哪儿了？
晚饭时间到了！

跳蚤的繁殖速度惊人，在短短的三周内，一对成虫可以在你的爱宠身上繁殖出千余只跳蚤。

跳蚤，真的？它
们把你折腾病还
不算完事吗？

1330年到1353年，一种被称为“黑死病”的鼠疫造成全球7,500多万人死亡。引发这场瘟疫的细菌（即鼠疫杆菌，译注）就是通过老鼠身上的跳蚤传播的。

虱子

几乎所有哺乳动物都有一个公敌，那就是虱子！虱子属于昆虫，大约有5,000种，它们体形很小，没有翅膀，都是体外寄生虫。虱子紧紧地附着在宿主的皮肤或毛发上，以吮吸宿主温热的血液或啃食死皮为生。我们人类是三种虱子的宿主，它们将虱卵产在我们的毛发里和衣服上。你感觉到痒了吗？

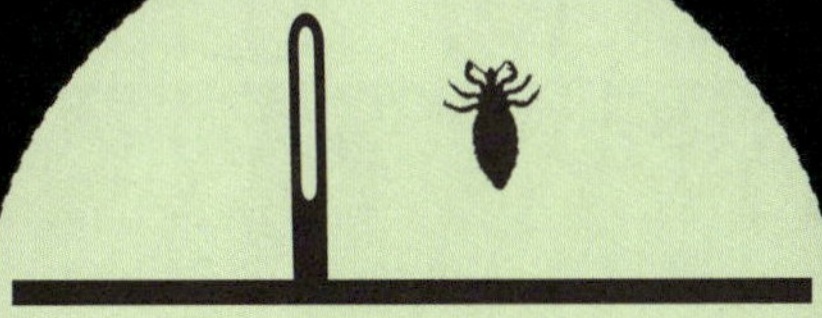

头虱

寿命：从虱卵产下到成虫死亡共30天

大小：体长2.5—3毫米

寄生虫

蝙蝠和鲸的身上不会寄生虱虫，但也有其他寄生虫。例如，鲸身上可能有长达2.5厘米的寄生甲壳动物。

虱卵

虱子对人类的折磨可能自古有之。当考古学家打开具有3,000年历史的埃及古墓时，居然在木乃伊上发现了虱卵。

皮肤上的爬虫

虱子主要有两种类型：一种吸食血液和其他体液；另一种啃食皮肤、羽毛、干结的血液和其他体表污垢。

生活在新几内亚岛的冠林鵙鹟，羽毛和皮肤都有毒。科学家认为，这可能是为了防范虱子而形成的一种适应性特征。

虱卵和虱子在远离宿主的条件下只能存活不超过24小时，所以虱子会用唾液将虱卵粘在宿主的毛发或羽毛上。

有种苍蝇叫做虱蝇，其行为很像虱子。有些种类的虱蝇没有翅膀，有的有小翅膀，它们都能紧紧地附着在宿主（比如狗或蝙蝠）的身上，吸吮宿主的血液。

在第一次世界大战期间，成千上万的士兵感染了“战壕热”。这是一种细菌性疾病，通过寄生在士兵衣服上的虱子进行传播，发病症状包括发热、头痛、腿痛等。当时士兵最喜欢的消遣活动就是挤衬衫上的虱子。

海蜘蛛

海蜘蛛有八条腿，模样和陆地上的蜘蛛有几分相似，但实际上是一种古老的海洋节肢动物，和蜘蛛并不是一家的。海蜘蛛遍布世界各地，无论是在浅水水域还是深海海沟，都有它们的足迹。海蜘蛛要么食腐，要么捕食，通过又长又尖的口鼻部从海葵等海底生命体中吸食营养物质。除了少数例外，它们都瘦得令人难以置信！

海蜘蛛
寿命：未知
大小：足展1—700毫米

无须呼吸

由于海蜘蛛通常都太瘦了，身体里没有空间安放肺或鳃，所以无法通过呼吸器官进行呼吸，而是通过身体直接从海水中吸收氧气。

小巧的肌肉

某些海蜘蛛的腿部肌肉极小，由单个细胞组成，只有在显微镜下才能观察到。

海底潜行者

海蜘蛛身体前端长有口鼻部和触肢，以及独特的、像钳子一样的器官——螯肢。当然，这和种类有关。并非所有的海蜘蛛都有全套装备。

海蜘蛛的腹部很细小，没有给内脏留出空间，所以它的内脏全长在腿里！海蜘蛛也没有鳃，它通过外骨骼摄取氧气。

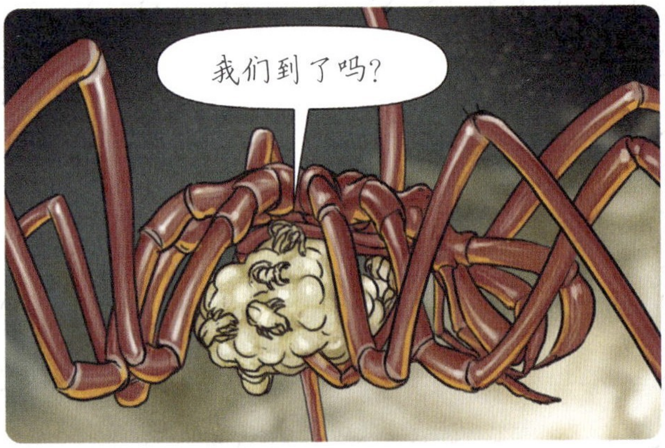

海蜘蛛的腿又长又细，在海底淤泥上跋涉时，这腿可是大有用武之地。在浅水区生活的海蜘蛛，腿部往往更为粗壮。

大多数海蜘蛛有一对携卵肢，通常折叠在身体下面。雄性海蜘蛛会用携卵肢来携带受精卵。

大多数海蜘蛛比蚊子还要小，但是在南极洲附近海域潜藏着巨型海蜘蛛，足展可达60厘米。它们和巨型蠕虫以及甲壳动物紧密地生活在一起。人们认为，寒冷的水中含氧量高，有助于巨型生物生存。

鼠妇

鼠妇共有3,000余种，它们属于甲壳动物，和虾、蟹等海洋生物是近亲。与其他甲壳动物不同的是，鼠妇生活在沙漠、高山等各种陆地栖息地上。鼠妇有个弱点：因为外骨骼会透水，除非找到黑暗、潮湿的藏身处（比如你家后院的花盆下），否则它们很快就会变干。

球鼠妇

寿命：通常为2年，有可能长达4年

大小：体长大约16毫米

一体两色

鼠妇分两步蜕皮：后半截身子先蜕皮，几天之后前半截身子再蜕皮。这就是为什么你有时会在同一只鼠妇身上看到两种颜色。

吃便便！

鼠妇吃自己的粪便来回收营养物质。它们不撒尿，而是释放出氨气。

虫界小坦克

哟，今天又是个
大热天。

还能再进去
一个吗？

一有机会，鼠妇就想方设法将扁平的身体挤入狭小的空间。生活在沙漠里的鼠妇会为自己、同伴和幼虫挖洞，将身体挤入洞里保持凉爽。

海蟑螂是鼠妇的近亲，生活在海滩的飞溅区。海蟑螂呼吸空气，藏身于潮湿的裂缝里，以腐烂的海草为食。

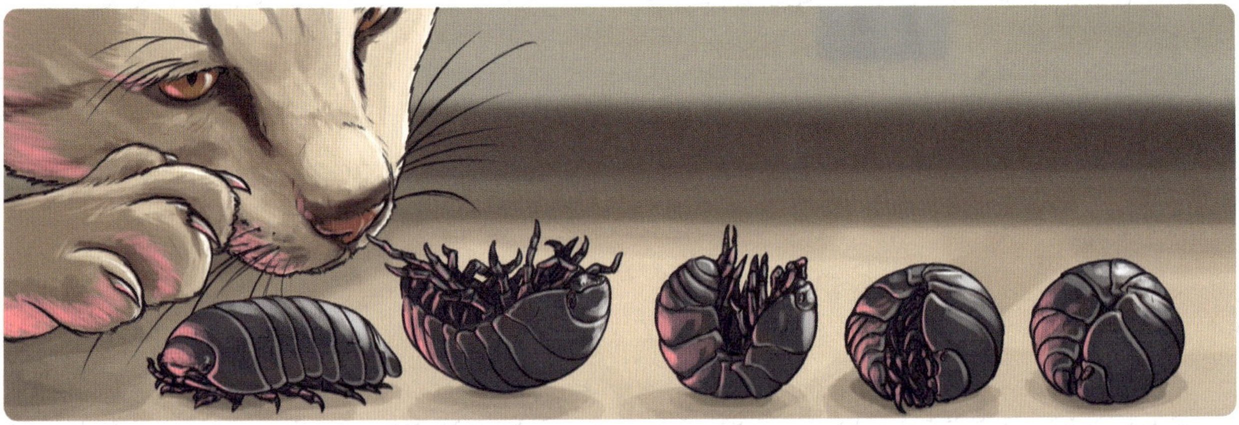

卷甲虫又称“球潮虫”或“卷布丁”，遇到危险时能像小犰狳那样把身体蜷缩起来。你可以在花园里找到它们。与鼠妇相比，卷甲虫的身体更圆。它们很容易与球马陆混为一谈，其实这两种动物相去甚远。

哦，痒死啦！
这叫掀背式，你
也试试呗！

但愿它们别让我
用鼻子顶足球。

等足目动物的幼虫被称为“缺肢幼体”，在母亲身体下部的育儿袋里孵化。上图左侧是一只球潮虫，它蜷着身体让幼虫出来。上图右侧是一只海蟑螂，它得抬高腹部才能让幼虫爬出来。

巨型等足目动物大王具足虫是鼠妇的远亲，体长可达36厘米，生活在寒冷的深水区。

天鹅绒虫

天鹅绒虫身体柔软，有多对足，摸上去毛茸茸的。它们不是蠕虫，而是一种非常古老的捕食者，下手毒辣，招招致命。天鹅绒虫生活在潮湿的落叶层中，在漆黑的环境下悄悄逼近猎物。它们先用触角轻触猎物身体，一旦断定这是个值得攻击的目标，便啪嗒一声从口乳突中喷出两股黏液，让猎物毫无招架之力。

天鹅绒虫

寿命：长达6年

大小：体长15—150毫米

多对足

世界范围内有大约180种天鹅绒虫，足的数量取决于种类，从13对到43对不等。它们的足上长有爪，用来在不平整的地面上增加额外的抓地力。

黑暗里的居民

天鹅绒虫喜欢阴暗的环境，部分原因是它们需要保持潮湿，防止变干。但如果过于潮湿，它们也会被淹死。

等级制度

有时好几条天鹅绒虫聚在一起进食，这种时候，谁先吃谁后吃，绝对不允许乱来。享有优先进食权的是最专横的雌虫。

古老的杀手

天鹅绒虫即有爪动物门动物，它们是非常古老的生物。人们发现的天鹅绒虫化石已有五亿年的历史了。

天鹅绒虫的嘴巴两侧有两个口乳突，就像可以移动的炮塔，能够将黏液喷射长达30厘米远。

天鹅绒虫的食谱中有鼠妇、蜘蛛、蟋蟀等。在悄悄靠近猎物之后，天鹅绒虫喷出黏液将猎物粘住，以防猎物逃跑。接下来，它张开强有力的颚部，用刀片一样的“牙齿”在猎物身体上咬出一个洞，注入唾液，将猎物的内脏变为可以吸食的黏糊糊的东西。

天鹅绒虫没有坚硬的外骨骼，只有一层柔软的表皮。为了生长，天鹅绒虫每隔几周就要蜕一次皮。

天鹅绒虫分不同种类，有卵生的，也有胎生的。对于胎生的种类来说，在经历了长达15个月的孕期后，天鹅绒虫妈妈直接产下幼虫，新生的天鹅绒虫一出生就能独立生活。

水熊虫

小小的水熊虫又称“缓步动物”，意为“行动缓慢的步行者”。它们穿行于潮湿的植物中间，用泵压式的口鼻部进食比自己更细小的动物或藻类。水熊虫是有史以来生命力最顽强的动物！在试验过程中，无论水熊虫被怎么折腾，煮沸、冷冻、加压、缺氧、遭受辐射，它们都能活下来。这些还不是最厉害的。水熊虫的看家本领是在无水环境下，进入所谓的“小桶”状态，把身体萎缩成干海绵状，慢慢等待环境变得湿润起来。有时候，这一等就是好几年。

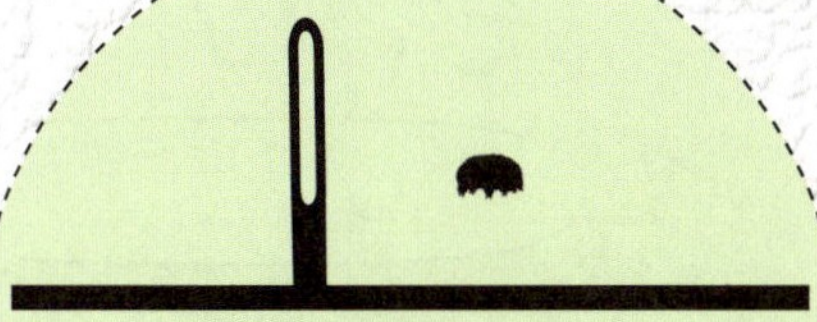

水熊虫

寿命：10年及以上

大小：体长0.1—1.5毫米

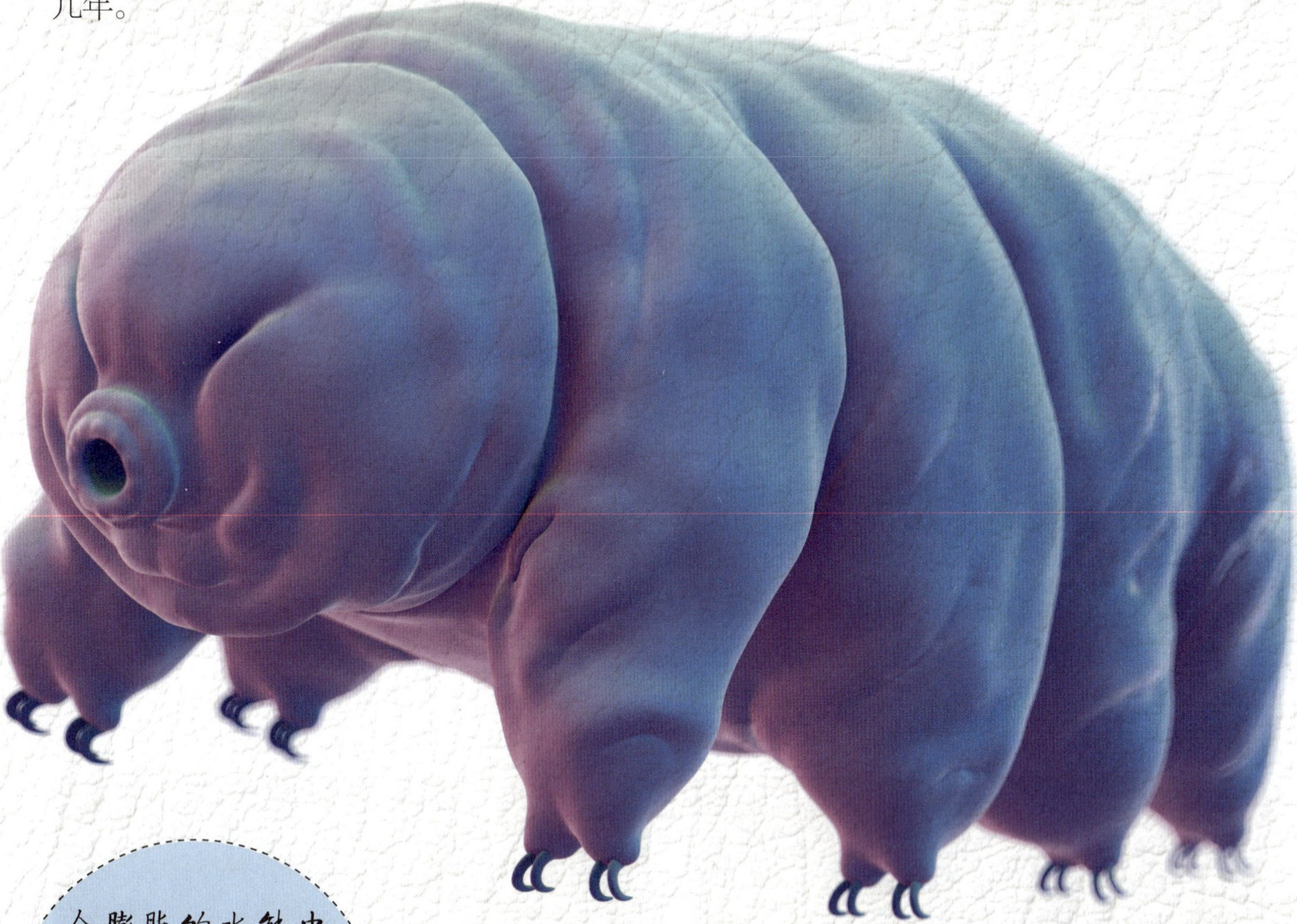

会膨胀的水熊虫

在缺氧时，水熊虫会进入“缺氧隐生”状态，它的身体会像上图中那样鼓起来，直到氧气的浓度恢复正常。

水熊虫是外星生物吗？

有人曾提出水熊虫可能是外星生物的观点。但水熊虫在缺氧状态下仅能存活几天，所以不太可能是天外来客。

抗高压能力

水熊虫能承受的压力是深海海沟最深处压力的六倍左右。

金刚不坏之身

水熊虫利用体液的压力让腿弯曲，这些腿的末端有尖尖的、可以弯曲的爪子。

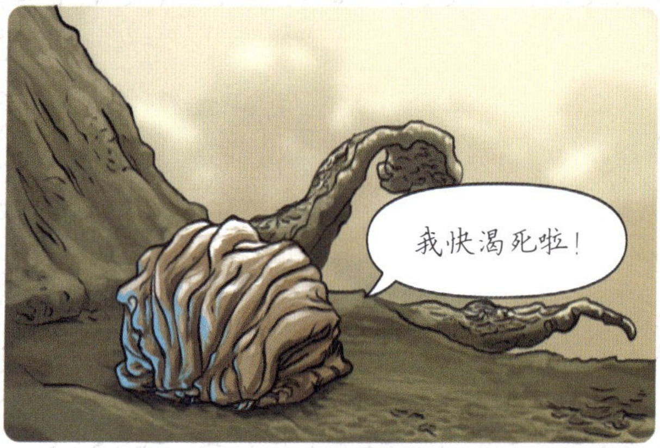

水熊虫在无水环境下会变干萎缩，进入"小桶"状态，体内一种叫做海藻糖的糖类会替代水分来维持细胞活性。

小桶状态下的水熊虫通常能挺过身体变干的无水期。风可能会将水熊虫和它们的卵带到新的地方，一旦再次变得湿润起来，它们就会开始繁殖。

试验表明，水熊虫耐得了高温，扛得住严寒，在-272℃—149℃的温度范围内都能够存活。这远远超过了人类的耐受范围！

2007年，欧洲航天局用火箭将小桶状态下的水熊虫和它们的卵带上太空，暴露在真空环境中，并且使其遭受足以杀死人类的高强度太阳辐射。十天后，这些水熊虫回到地球，经过复水处理后，它们中有三分之二的个体毫发无损地活了下来。

蜱虫

鼠妇

马陆

在本书中，你会看到关于一些寄生虫的介绍，它们有的寄生于我们的体表，比如跳蚤和蜱虫，也有一两种寄生于我们体内。另外，你还会了解到一些能惹出大麻烦的生命形态，比如鼠疫杆菌。这种细菌非常小，因此可以在微小的寄生虫体内传播，而它又极其致命，甚至曾改变人类历史的进程。小虫子主宰世界！

小虫子，大本领

不要被动物的个头所欺骗。最强大的生物中有些正是个头最小的小不点儿！在本书中，让我们一起来了解一些彪悍的小虫子。在这些虫子当中，有的我们在家就能看到，比如鼠妇、蜈蚣；还有的太小，需要在显微镜下观察。但它们有一个共同点——个个身怀绝技，比如惊人的自卫术、跑得快、跳得高、能够在极端环境中存活下来，等等。

水熊虫

跳蚤

不幸的是，当我们进入微观世界后，会发现很多令人讨厌的小虫子。跳蚤和蠕虫这样的寄生虫生活在宿主动物的体表或体内，通常会引起宿主不适。动物中高达50%的种类在某种程度上都是寄生的。对于寄生虫而言，毕竟寄生生活可以让食物获取更加方便，也可以提供高级别的保护，这种日子真是够美的！

目 录

最强大的生物中有些个头却是最小的！有些小虫子就生活在我们的家里，比如鼠妇和蜈蚣；还有些特别小，只能用显微镜才看得清楚。在这本书中，你将了解：

- 能够连续跳跃 30,000 次的跳蚤
- 能够产生氰化物的马陆
- 从具有 3,000 年历史的木乃伊上找到的虱卵

……

还有更多关于生命力顽强的小虫子的奥秘。你将在书中与这些小虫子面对面，认识那些坚不可摧的水熊虫、内脏长在腿里的海蜘蛛，以及用足尖感受气味的蜱虫等。小虫子无处不在！

番茄博士双语百科

不可思议的虫子王国

小虫子，大本领

马特·特纳（英）著
圣地亚哥·卡列（英）绘
丛岚 译

外语教学与研究出版社
FOREIGN LANGUAGE TEACHING AND RESEARCH PRESS
北京 BEIJING